PULSE OF THE PLANET No.2:
On Cosmic Energy, Acupuncture Energy, A-Bombs & Earthquakes, and Wilhelm Reich's Orgone Accumulator

The occasional research report and journal of the
Orgone Biophysical Research Laboratory, Inc.

Editor: James DeMeo, Ph.D.
Production Assistant: Theirrie Cook, BA

CONTENTS: **Page**

Distributed by Natural Energy Works / Ingram / Lightning Source
www.naturalenergyworks.net   ISBN: 978-0989139052   ISSN: 1041-6773   150626

Editor's Page

This summer has been incredibly busy for workers at the Orgone Biophysical Research Laboratory (OBRL). Three workshops, five major field expeditions into the deserts of Arizona, invited lectures at several different Conferences and Symposia, including in West Germany and Japan, and demands for articles appearing in other publications, all have taxed our time and resources. Accordingly, we found ourselves getting farther and farther behind in the preparation of the *Pulse of the Planet*. For several reasons, it was decided to limit publication of the *Pulse* to only twice per year. As our primary mission is *research*, and as we do not receive enough money from subscriptions or donations to justify either paying ourselves a salary, or hiring on staff, this was a necessary step. However, as you will note, this issue of the *Pulse* is almost double the size of the first issue. We now cover a full six months of weather and geophysical data in the "Climate Features and Unusual Phenomena" section, and we have expanded our "Notes" considerably. A new section on "Conference Reports" has also been included. Even though the *Pulse* will appear only twice a year, it will not compromise on quality or substance, and will remain one of the best information sources on new developments in orgone biophysical research.

The Laboratory is building an important international network of contacts with researchers who are confirming the existence of an energy in space, in the human body, or in the Earth. Among almost all of these people, we are finding that the discoveries of Reich are well respected. From a Canadian professor, Dr. Gary Whiteford, the *Pulse* presents evidence on the link between underground nuclear bomb testing and earthquakes, in what is the first formal publication of his data in an American publication. We likewise present a summary of important research materials on the orgone energy accumulator by researchers from Germany. The Diploma Thesis presented by Stefan Müschenich and Rainer Gebauer from the University of Marburg was the first double-blind, controlled study of the accumulator, and it confirmed many of Reich's original findings. Additionally included is a fascinating paper from Prof. Dr. Bernd Senf, who experimentally explored the relationship between orgone energy, and the energy routinely used by acupuncturists. A paper by this Editor provides a broad overview of research by various natural scientists who have stumbled upon the orgone energy in their own experiments. The evidence to support Reich was never ambiguous or weak, but it is made even stronger by these findings, which we are proud to present.

James DeMeo, Ph.D.
El Cerrito, California, September 1989

Postscript, to the 2015 Reprint

In 2015, it was decided to reprint the long unavailable back issues of *Pulse of the Planet*, for several reasons. Firstly we were not satisfied providing low-quality photocopy versions of the out-of-print *Pulse* issues. Secondly, the old back issues retained good content value, and thirdly they documented a lot of material useful for historical purposes. While there are a few things in this *Pulse* No. 2 that I would have worded differently today, or perhaps not have included, in the main, the findings and notations contained herein have withstood the test of time, and retain their value.

This particular issue presented new findings in orgonomic science of foundational value for 1990, and will continue to be entirely new and eye-opening for those who only recently learned about Wilhelm Reich. It will remind people of the intensive controversy surrounding atomic bomb tests and geophysical phenomena, following a trend set in the first issue of *Pulse* in Spring 1989, and continued in subsequent issues. We were also happy to introduce to the English speaking world the findings of German scientists on the orgone energy accumulator, and on orgone acupuncture. As noted in the original Editor's Page, this Issue No.2 also detailed an incredibly busy period for me personally. In the two years following my 1988 abandonment of a prior role as university professor, for a new role as independent research scientist, I worked towards establishing a research laboratory and educational center on the West Coast USA. That goal was only finally achieved in Summer of 1994, with purchase of rural land in Oregon near to Ashland. This issue of *Pulse* also carried forward the intensive effort at global monitoring and mapping of geophysical, atmospheric, environmental and biological extremes and anomalies, in efforts to better understand the totality of bioenergetic coherence in nature. Our desire was to provide a weekly summary version of what the *Center for Short-Lived Phenomena* of the Smithsonian Institute originally provided, which had been shortsightedly defunded in 1974.

Pulse was initially conceived as a quarterly journal, which was far too ambitious given our meager financial resources. In fact, after this Issue No.2, *Pulse* became an irregular *Occasional Report of the OBRL*. This high-quality reprint edition, available for the first time in over 20 years, is unchanged from the original, aside from the addition of this Postscript, and the occasional correction or strikethrough of expired addresses and other contact information.

James DeMeo, Ph.D.
Orgone Biophysical Research Lab
Ashland, Oregon, USA
Summer 2015

The Orgone Energy Continuum: Some Old and New Evidence

James DeMeo, Ph.D.*

This paper was first presented to the Eleventh International Congress of Biometeorology, at Purdue University in1987, and later in 1989 at the First International Congress on Geo-Cosmic Relations, in Amsterdam, The Netherlands. See the "Conference Reports" section of the *Pulse* for more details on these events.

From 1934 to 1957, Wilhelm Reich published a series of experimental reports demonstrating the existence of a specific biologically and meteorologically active energy continuum at work within the atmospheric ocean, and in high vacuum; he called this energy the *orgone*. Reich developed unique experimental devices, notably a layered organic/metallic energy-accumulating enclosure, which he called the orgone energy *accumulator* (1), and a special water-grounded pipe antenna for affecting clouds and weather patterns from a distance, called the *cloudbuster* (2). His research findings on these questions span over 30 years, and there has been an additional 30 years of corroborative work by others, undertaken since his death in 1957 (3); this paper will therefore be confined to only a few aspects of Reich's works that have been investigated and confirmed by the author. Additionally, a few parallel discoveries by other scientists will be reviewed, and contrasted to those of Reich.

Reich's most crucial experiments and observations have been verified, duplicated, and strengthened in both Europe and the USA. In controlled studies, orgone energy accumulators, which resemble the Faraday cage in some respects, have been found to have specific influences upon air temperature, humidity, and electroscopical discharge rate in a manner dependent upon local meteorology. A stimulating effect upon plant growth, wound healing, and general physiology has also been documented (3,4). My own studies have verified such effects upon seeds sprouted inside the orgone accumulator; mung beans grown in water inside an orgone accumulator have been observed to sprout to lengths as much as six times those grown inside control enclosures. Figure 1 expresses the results of a past seed sprouting study in graphic form (5). Controlled studies have also shown that the growth-enhancement effect occurs when seeds are charged in a dry condition, and subsequently planted in garden plots (6). The accumulator used by the author in this experiment was a cube shape, composed of ten alternating layers of steel wool and acrylic plastic sheeting, with a final interior lining of galvanized steel plate, and a final exterior sheet of organic celotex. Accumulators must be constructed with the metals on the inside and organic, non-metals on the outside, though they can be layered for added charging strength. For biological experiments, steel or galvanized steel must be used, as copper or aluminum impart a toxic effect. Also, accumulators must be strictly isolated from any kind of electromagnetic devices or nuclear ionization materials; these impart a disturbing, toxic quality to the energy. Control enclosures, by comparison, employ only organic, non-metallic materials.

My own work subsequently demonstrated an evaporation-suppression effect at work inside the orgone accumulator (7); the accumulator will evaporate less water than a control enclosure, even when significant mixing of external environmental air is allowed to occur. The accumulator suppresses evaporation most strongly on bright sunny days, when its charge is strongest. This increased charge inside the accumulator on bright sunny days is generally sensible by the average individual as a radiating warmth or tingling sensation that occurs at the skin surface. The accumulation effect is greatly minimized, or vanishes entirely, on rainy overcast days. Figure 2 graphically displays the changing values of the evaporation suppression effect (evaporation inside an orgone accumulator minus evaporation inside a control enclosure). This effect has nothing to do with air temperature, as both the accumulator and control enclosures were kept in shaded, equal-temperature environments.

*   Director of Research, Orgone Biophysical Research Lab., Ashland, Oregon, USA   demeo@orgonelab.org

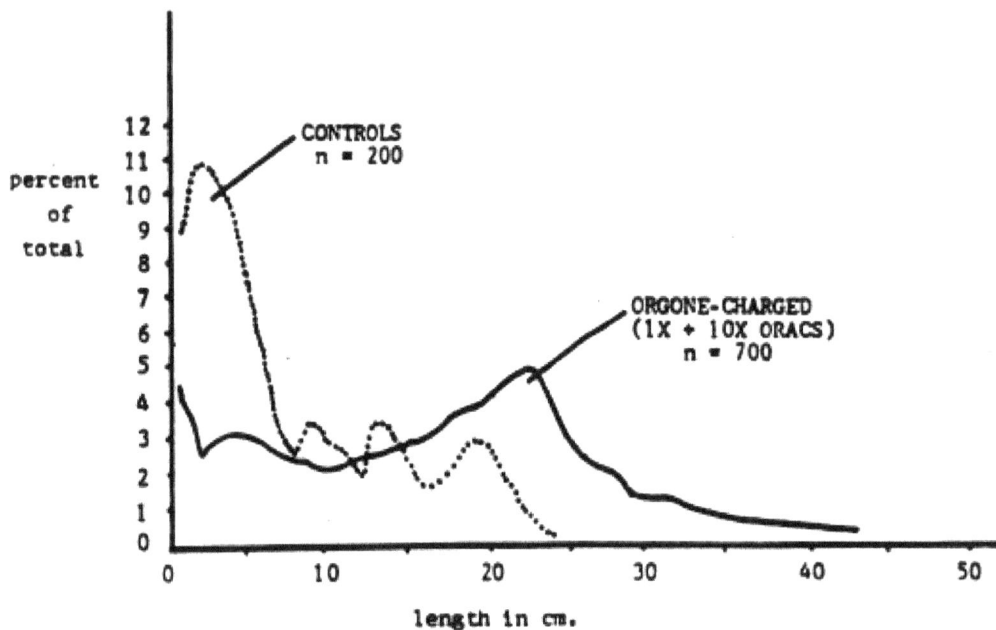

Figure 1: Frequency distribution of the length of mung beans sprouted within a galvanized steel and celotex orgone energy accumulator (solid line) and a metal-free cardboard control enclosure (dotted line). The orgone charged sprouts were significantly longer than the controls (5).

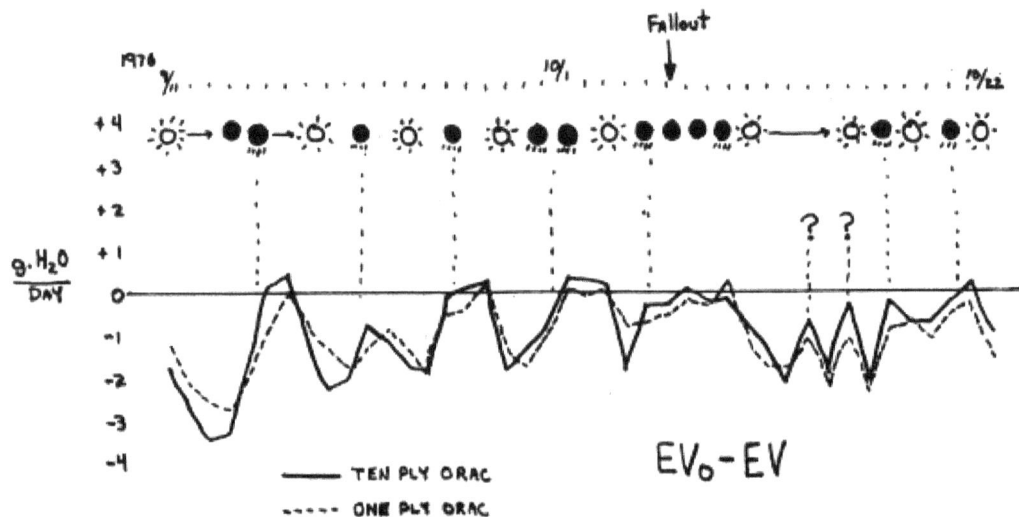

Figure 2: Evaporation differential, grams of water evaporated per day within a galvanized steel and celotex orgone energy accumulator *minus* grams of water evaporated within a metal-free cardboard control enclosure. Note the suppression of evaporation within the accumulator on bright sunny days; the effect diminishes or vanishes on rainy or overcast days (7).

The waxing and waning of the orgonotic charge in the evaporation experiment is very similar to other curves expressed in the accumulator *temperature differential* (To-T) and *electroscopical discharge rate differential*, all of which indicate that the interior space of the accumulator possesses a stronger energetic density or charge (3). The observations on plant growth enhancement, physiological stimulation and wound healing suggest that the accumulator actually accumulates some kind of environmental energy which is likewise fundamental to life processes. For these and other reasons, Reich believed the orgone energy was a *life energy*, satisfying the conceptual requirements of both a "vital force" and "aether" (8). Reich also believed that the increase and decrease of the orgone charge in the accumulator, timed to weather events, was indicative of an energetic *atmospheric pulsation*.

In his later years, Reich focused upon atmospheric questions, and he developed a unique instrument, the cloudbuster, which he demonstrated could restore atmospheric pulsation, clouds and rains, to regions suffering from drought or aridity. The cloudbuster instrument, which must be partly grounded in moving, unpolluted water, has been investigated by the author; the instrument has a demonstrated capacity to influence cloud dynamics over very large regions, and has been used to break droughts and bring rains in various parts of the world (3).

Figure 3, for example, summarizes the results of 12 tests to bring rain in Kansas during 1977-1978. That study demonstrated an unusual peaking of precipitation within hours after onset of cloudbuster tests, which occurred during normal climatic conditions, without drought tendencies (9). Other cloudbusting experiments have been undertaken by the author during droughty conditions (10). Those experiments have indicated that strong atmospheric pulsation, and a return of regular rains, will often develop in drought regions within roughly 48 hours after onset of cloudbusting operations. Approximately 80% of the author's 35 cloudbusting experiments, and 80% of the 18 cloudbusting experiments performed under droughty or arid conditions, have been followed by increased atmospheric pulsation, cloud cover, and rains. More recently, a series of cloudbusting experiments undertaken near Yuma, Arizona (the driest part of the USA) have indicated that even the highly immobilized, stag-

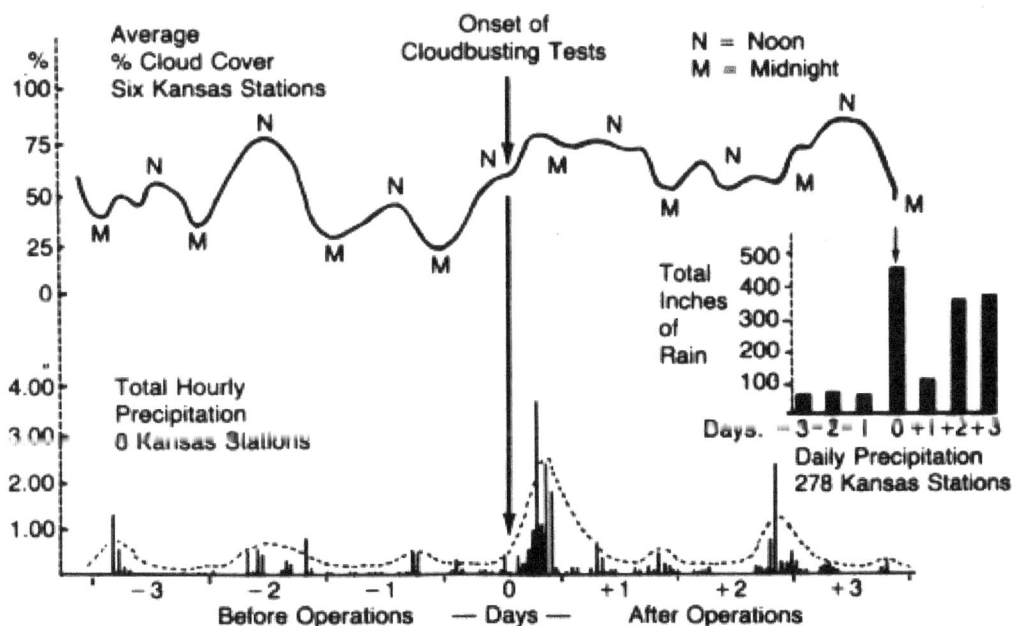

Figure 3: Summary of weekly cloud cover and precipitation data in the State of Kansas, for 12 cloudbusting operations to bring rain, 1977-1978, non-drought conditions. National Weather Service data is used. The arrow marks the time of onset of cloudbusting operations, after which a clear increase in percent cloud cover and rainfall occurs over Kansas (9).

OROP Arizona Weather, August 1988

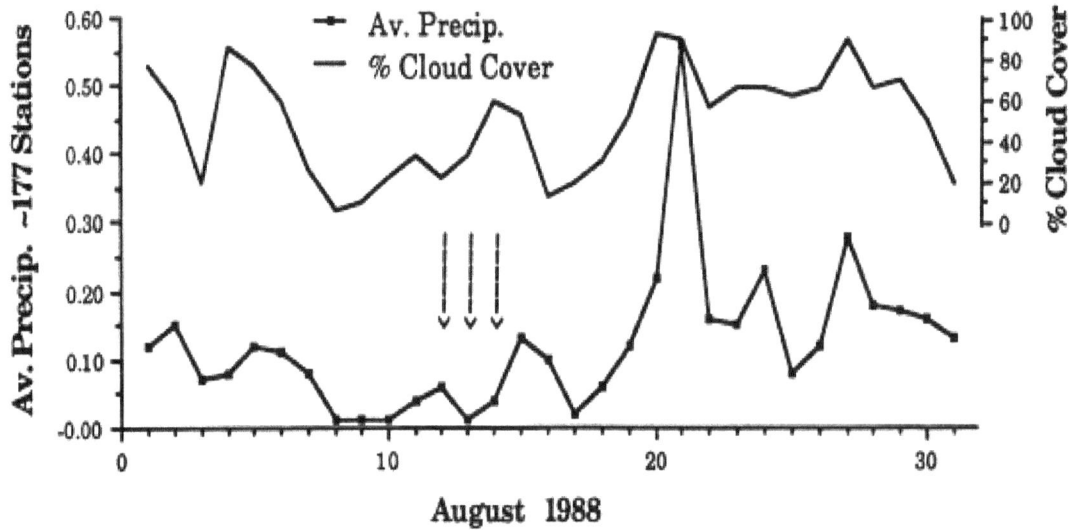

Figure 4: Summary of monthly cloud cover and precipitation data in the State of Arizona, August 1988. Arid desert conditions. National Weather Service data is used. Cloudbusting operations ran from August 12-14, as noted by the arrows. Rains dramatically increased in the week following operations. Note the high percent cloud cover, but low rainfall for the first half of the month (11).

OROP Washington Weather, September 1988

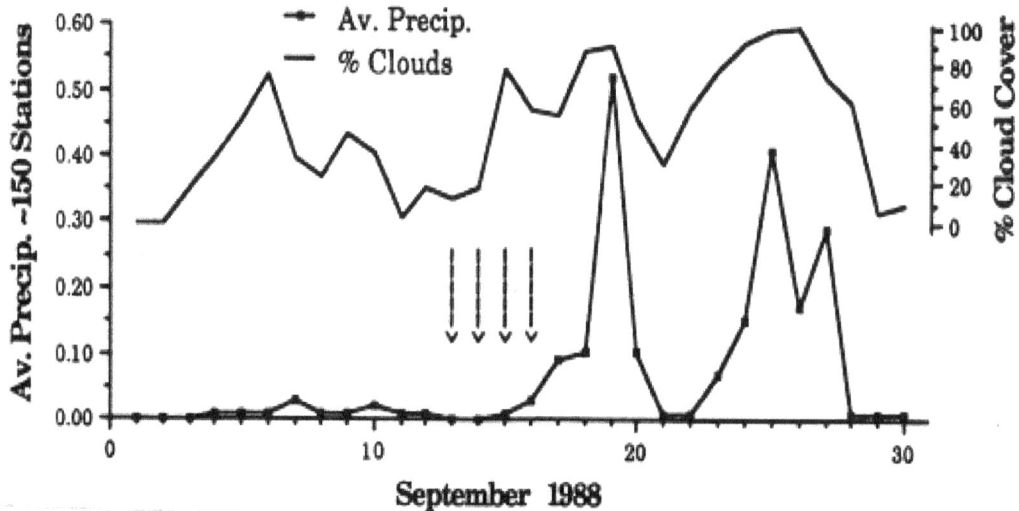

Figure 5: Summary of monthly cloud cover and precipitation data in the State of Washington, September 1988. Extreme drought conditions. National Weather service data. Cloudbusting operations ran from September 13-16, as noted by the arrows. Rains dramatically increased in the days following the operations, and persisted thereafter in a pulsatory manner. Note the high percent cloud cover, but low rainfall for the first half of the month (11).

nant, overheated and cloud-free atmosphere of the harsh desert may respond to the cloudbuster's influence. Figures 4 and 5 show rainfall data for one cloudbuster operation undertaken under desert conditions, and one during extreme drought conditions, respectively (11).

The kinds of experimental results summarized above are entirely inexplicable from the classical point of view, but they do support Reich's assertion regarding the existence of a water-influencing, biological-meteorological-cosmic energy continuum. Evidence suggests that other scientists working completely independent of Reich have measured or detected similar phenomena. Primary on this list is Giorgio Piccardi, who demonstrated that metal enclosures or shielding would affect physio-chemical phenomena in a manner that revealed cosmic and meteorological factors (12). Piccardi's observations on the effects of metallic chambers upon experimental results, and specifically on the phenomenon of water activation, parallel those of Reich on the effects of layered organic/metallic chambers (orgone accumulators), and the capacity of water to absorb the atmospheric orgone energy. This latter principle is employed most effectively in the cloudbuster. While Piccardi attempted to explain the results of his experiments within the framework of classical electromagnetic theory, Reich's findings suggest that the effect is the result of a more fundamental natural force. At least one of Piccardi's co-workers broke with traditional electromagnetic concepts in an attempt to explain the phenomena, proposing the existence of a specific weather-radiation (13).

Other scientists have measured unusual natural phenomena which are conceptually similar to Reich's orgone, involving the use of metallic enclosures for detection of the phenomena of interest. A listing of these kinds of unusual natural phenomena can be given:

BIOENERGETIC FACTORS:

Harold Saxon Burr (14), using the same kind of sensitive millivoltmeter that led Reich towards the discovery of the orgone, measured fluctuating *electrodynamic fields* within given environments that determined the electrical activity and behavior of organisms and objects. His findings were reminiscent of the fluctuating atmospheric pulsation detected within Reich's accumulators, and within Piccardi's enclosures. Like Reich, Burr also found that the strength of the charge of tissues was a determinant of health; bioelectrical charge had a determining influence upon both tissue structure and whole system behavior.

Rupert Sheldrake (15) recently wrote about the participation of a hypothetical energetic *morphogenetic field* in the structuring and ordering of matter, life, and hereditary processes.

Bjorn Nordenstrom (16) studied the unusual "x-ray ghost" phenomenon, as well as the bioelectrical aspects of diseases, to conclude that the human body possessed a separate circulatory system for bioenergy.

Robert O. Becker (17) made a similar analysis of bioelectrical phenomena accompanying wound and bone healing, and developed practical applications for the bioelectrical stimulation of the healing process, to include the stimulated regrowth of amputated limbs of mammals.

Frank Brown (18) similarly identified various cosmic-spatial and meteorological components at work in the rhythmical behavior of living systems. His work often involved the use of metallic enclosures.

Thelma Moss (19) studied the phenomenon of "Kirlian" electrophotography, making photos of the energy fields of people, plants, and various objects. Her phantom leaf images — showing a whole leaf on photos where part of the leaf was previously cut away — refuted the assertion that the effect was wholly one of an electrical nature, and she at last developed a non-electrical method for making energy field photos, involving use of an orgone accumulator to enhance the energy fields of the object to be photographed.

Jacques Benveniste (20) recently provided experimental corroboration for the principle of homeopathic dilutions, a non-molecular principle in which energetic charge with specific informational characteristics is maintained in solutions of very high dilution strengths.

Louis Kervran (21) provided detailed experimental evidence for the transmutation of non-radioactive elements by living creatures (microbes and mammals) as a part of ordinary metabolism. He believed that the transmutations were driven by some form of biological or vital energy.

GEOCOSMIC FACTORS:

Dayton Miller (22) measured a *dynamic aether drift* using a light-beam interferometer similar to, but more sensitive than the ones previously used by his teachers, Michelson and Morley. Contrary to the popular myths regarding the non-existence of the aether, Miller demonstrated that the aether drift was real and dynamic, in motion with the Earth, more active at higher altitudes due to entrainment effects, and capable of being reflected by metals. These latter points are very much in keeping with Reich's and Piccardi's findings.

Halton Arp (23) continues to make photographs of deep-space objects that are energetically intertwined and structurally connected, but which have vastly different red-shifts. His finding undermine the use of redshifting as indicators of astrophysical distance, and also the big-bang theory, but do support the concept of a universe rich in a light-affecting, aether-like substrate.

Hannes Alfven (24) has argued for the existence of cosmic plasmas in deep space.

Various laboratory workers (25) have demonstrated the non-constancy of radioactive decay "constants", and unusual non-linear continuum effects in nuclear decay processes. These findings may be interpreted as evidence that the nucleus is affected in a powerful manner by outside influences of an unknown nature.

Based upon the theoretical abundances of all the neutrinos created since the beginning of time, **Paul Dirac**, **Louis deBroglie**, and others (26) theorized that an aether or *neutrino sea* must exist.

All of the above researchers have demonstrated either the importance of cosmic spatial/environmental factors to experimental results, or the necessity for the existence of a biologically and meteorologically active aether-like energy continuum. In a few cases, such an energetic force was directly detected, measured, and identified. While all of the above workers have made fundamental new discoveries which suggest or require the existence of unusual new forces, the works of Reich, Piccardi, Burr, Brown, and Miller provide the clearest evidence as to the nature of the phenomena. The evidence strongly suggests the existence of a cosmic-atmospheric-biological energy at work in the natural world, which exerts influences on both living and non-living systems through energetic and spatial influences correlated with the helicoidal movement of the Earth around the Sun, and the Sun's movement through the galaxy.

Importantly, each of the above workers came to their conclusions entirely independent of each other, and the evidence they have gathered supports the findings of each other in a fundamental manner, even though differences in theory and approach exist. The energetic forces they measured have cosmic components which are difficult or impossible to shield, and metallic shields or screens often appear to *amplify* the phenomenon. Water is a very important agent for absorbing or transmitting the energetic influences in question. Taken as a whole, the above discoveries provide powerful evidence for a new natural phenomenon, which has been called different names by different researchers, depending upon their research discipline. While some of the above workers have chosen to retain classical electromagnetic and biological theorems to explain the phenomena, many have abandoned those theorems in favor of models that more accurately accommodate the results of their experiments. The classical theories of "empty space" and "biochemistry", and the inviolability of the Second Law of Thermodynamics are most powerfully undermined by these findings. Burr's electrodynamic field theory and Miller's dynamic aether

drift theory are, for example, irreconcilable with much of existing theory. Reich's theory on the orgone energy is the most radical departure from conventional views, but in the author's view, it is a most dynamic, comprehensive, and powerful theorem, which is congruent with observed, experimentally demonstrated fact, and therefore worthy of serious attention.

REFERENCES:
1. Reich, W.: *Discovery of the Orgone, Vol. 2: The Cancer Biopathy*, Orgone Institute Press, NY; 1948: *The Orgone Energy Accumulator, Its Scientific and Medical Use*, Orgone Institute Press, Rangeley, ME, 1949.
2. Reich, W.: "Dor Removal and Cloudbusting", *Orgone Energy Bulletin,* IV(4):171-182, 1952.
3. DeMeo, J.: *Bibliography on Orgone Biophysics*: 1935 - 1986, Natural Energy Works, PO Box 864, El Cerrito, CA 94530 USA, 1986.
4. Muschenich, S. & Gebauer, R.: *Der Reichsche Orgonakkumulator*, Nexus Press, Berlin, 1987.
5. DeMeo, J.: "Seed Sprouting Inside the Orgone Accumulator", *J. Orgonomy,* 12(2):253-258, 1978.
6. Espanca, J.: "The Effect of Orgone on Plant Life, Parts 1 - 7", *Offshoots of Orgonomy*, #3 to #12, 1981 to 1986.
7. DeMeo, J.: "Water Evaporation Inside the Orgone Accumulator", *J. Orgonomy,* 14(2):171-175, 1980.
8. Reich, W.: *Ether, God & Devil*, Farrar, Straus & Giroux, NY; 1973; *Cosmic Superimposition*, Wilhelm Reich Foundation, Rangeley, ME, 1951.
9. DeMeo, J.: *Preliminary Analysis of Changes in* Kansas *Weather Coincidental to Experimental Operations with a Reich Cloudbuster*, Thesis, University of Kansas; reprinted by Natural Energy Works, PO Box 864, El Cerrito, CA 94530 USA, 1979.
10. DeMeo, J.: "Field Experiments with the Reich Cloudbuster: 1977 to 1983", *J. Orgonomy,* 19(1):57-79, 1985; DeMeo, J. & Morris, R.: "Preliminary Report on a Cloudbusting Experiment in the Southeastern Drought Region, August 1986", *Southeastern Drought Symposium Proceedings*, March 4-5, 1987, Columbia, SC., South Carolina State Climatology Office Publication G-30, pp.80-87, 1987.
11. DeMeo, J.: "CORE Report #20: Breaking the Drought Barriers in the Southwest and Northwest USA", *J. Orgonomy,* 23(1):97-125, 1989.
12. Piccardi, G.: *Chemical Basis of Medical Climatology*, Charles Thomas, Springfield, IL., 1962.
13. Bortels, J.: "Die Hypothetische Wetterstrahlung als vermutliches Agens Kosmo-Meteoro-Biologischer Reaktionen", *Wissenschaftliche Seitschrift der Humboldt-Universitat zu Berlin,* VI:115-124, 1956.
14. Burr, H.S.: *Blueprint for Immortality*, Neville Spearman, London, 1971; cf. Ravitz, L.: "History, Measurement, Applicability of Periodic Changes in the Electromagnetic Field in Health and Disease", *Annals*, NY Acad. Science, 98:1144-1201, 1962.

15. Sheldrake, R.: *A New Science of Life*, The Hypothesis of Causative Formation, J. P. Tarcher, Los Angeles, 1981.

16. Nordenstrom, B. : *Biologically Closed Electric Circuits*: *Clinical, Experimental and Theoretical Evidence for an Additional Circulatory System*, Nordic Medical Publications, Stockholm, Sweden, 1983.

17. Becker R. & Selden, G. : *The Body Electric:* Electro*magnetism and the Foundation of Life*, Wm. Morrow, NY, 1985.

18. Brown, F.: "Evidence for External Timing in Biological Clocks", in *An Introduction to Biological Rhythms*, J. Palmer, ed., Academic Press, NY, 1975.

19. Moss, T. : *The Body Electric: A Personal Journey Into the Mysteries of Parapsychological Research, Bioenergy, and Kirlian Photography*, J. P. Tarcher, Los Angeles, 1979.

20. Davenas, E. & Benveniste, E.: *Nature*, 333:832, 1988.

21. Kervran, L.: *Biological Transmutations*, Beekman Press, Woodstock, NY, 1980.

22. Miller, D.: "The Ether-Drift Experiment and the Determination of the Absolute Motion of the Earth", *Reviews* of *Modern Physics*, 5:203-242, 1933.

23. Arp, H., et al: *The Redshift Controversy*, W.A. Benjamin, Reading, MA, 1973; *Quasars, Redshifts, and Controversies,* Interstellar Media, Berkeley, CA, 1987.

24. Alfven, H.: *Cosmic Plasmas*, Kluwer, Boston, MA, 1981.

25. Berkson, J.: "Examination of Randomness of Alpha Particle Emissions", *Research Papers in Statistics*, F.N. David, ed., Wiley, NY, 1966; Emery, G.: "Perturbation of Nuclear Decay Rates", in *Annual Review of Nuclear Science*, Annual Reviews, Palo Alto, CA, 1972; Anderson, J. and Spangler, G.: "Serial Statistics: Is Radioactive Decay Random?", *J. Physical Chemistry*, 77:3114-3121, 1973.

26. Dirac, P.: "Is There An Ether?", *Nature*, 162:906, 1951; deBroglie, L.: *Non-Linear Quantum Mechanics*, Elsevier, NY, 1960; Dudley, H.: *New Principles in Quantum Mechanics*, Exposition University Press, NY, 1959; Dudley, H.: *Morality of Nuclear Planning*, Kronos Press, Glassboro, NJ, 1976.

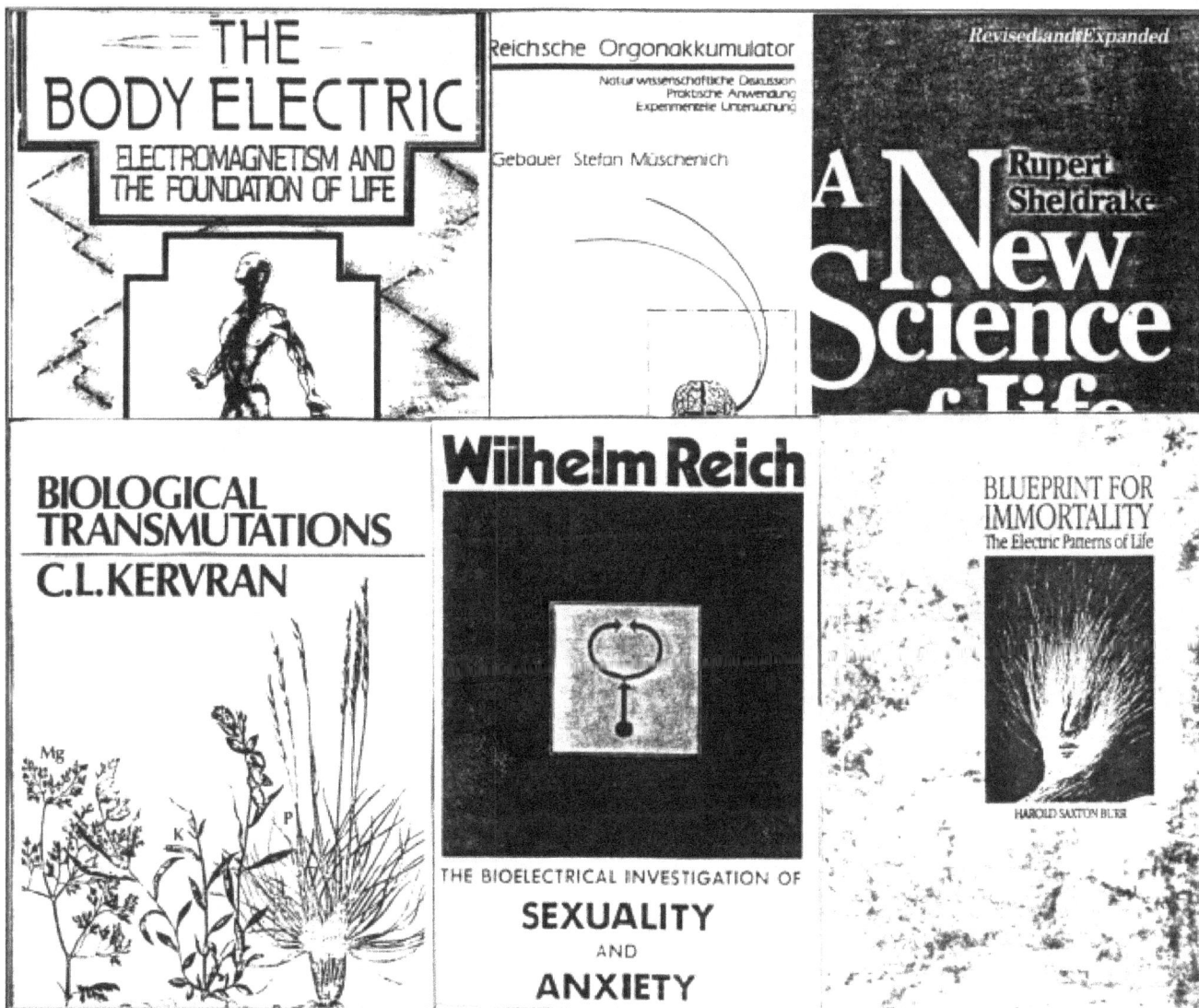

Editor's Preface to Whiteford Paper:

In 1988, the Pulse received a translation of a report in a popular German magazine about a Canadian professor who had found a correlation between nuclear bomb tests and earthquake patterns. We finally located Dr. Gary Whiteford, at the University of New Brunswick, who filled us in with the details, and offered the following paper to the Pulse. There had been a near black-out regarding his research in the USA, and no academic journal appeared willing to publish his controversial research findings, which he openly admits are preliminary and conditional. The opposition to his work focuses upon the assumption that there is "no mechanism" by which such long-distance, and oftentimes delayed geophysical responses could occur following detonation of a nuclear bomb. However, Whiteford's investigation is very important, given the incredibly dangerous consequences to the planet should the observed patterns actually be correct. Readers of the Pulse will recall a previous paper on this question published in the last issue, by Yoshio Kato, a Japanese researcher, whose study sat unpublished for more than a decade! To refresh the reader, Kato observed that underground nuclear tests were having an effect upon the rotational dynamics of Earth, and also upon earthquake patterns. He argued also that dramatic increases in upper atmospheric temperatures followed underground tests, which even more strikingly violated the question of "mechanism", and so his paper sat unpublished for years. Now comes a recent study by the Lawrence Livermore National Laboratory, linking atmospheric nuclear bomb tests to the deterioration of the stratospheric ozone layer. This study does not directly corroborate Kato's findings, which focused only upon underground nuclear bomb tests. However, the man who undertook the Livermore Labs research acknowledged to us that only a few years ago the same argument of "no mechanism" could equally have been applied to the question of upper atmospheric changes from atmospheric nuclear bomb testing. While we respect the right of anyone to voice disagreement with Kato and Whiteford, it is amazing how often this disagreement is made purely as a defense of mechanistic atomic theory, the same inadequate theory which could not predict the deadly consequences of nuclear technology. Mechanistic atomic theory, which denies the evidence for an excitable vital energy principle in living creatures, in the atmosphere, in space and in the Earth, also provides the theoretical structure and intellectual ammunition for the anti-nature world view, which underlies the big global mess we are currently facing! That Dr. Whiteford (and others) are identifying natural processes that require a more functional, energetic theory to explain them should not be cause for either alarm, or an auto-da-fe.

Bomb Tests Nuked Ozone

Washington (FOE) - The Lawrence Livermore National Laboratory has concluded that above-ground testing of massive nuclear warheads during the 1950s and early 60s seriously damaged the ozone layer. Using computer models and atmospheric data available for the period, they found that detonation of 15 to 58 megaton hydrogen bombs caused as much as a 5 percent loss of ozone over the central United States, and cut Arctic Ozone by up to 12 percent. Donald Wuebbles, a scientist with Livermore Labs said that "the initial effect was at the high latitudes" where the largest warheads were exploded by the Soviet Union. "We found significant decreases in ozone over North America", he added. The bomb explosions lofted atomic fireballs into the stratosphere. The intense heat produced short-lived nitrogen ions which initiated a catalytic cycle splitting ozone into oxygen and nitrogen oxides.

The United States and the Soviet Union agreed in 1963 to limit nuclear tests when they signed the Partial Test Ban Treaty. Elimination of this source of nitrogen compounds in the stratosphere allowed ozone levels to gradually recover over the rest of the decade. (From *Atmosphere*, Winter 1989, Friends of the Earth International, 218 'D' Street, SE, Washington, DC 20003)

Earthquakes and Nuclear Testing: Dangerous Patterns and Trends

Gary T. Whiteford, Ph.D. *

Presented to the Second International Conference on the United Nations and World Peace, Seattle, Washington, 14 April 1989.

Introduction

This paper is an attempt to understand distributions, patterns, and directions of large earthquakes of Richter Magnitude (M) >= 6 since 1900. Secondly, attempts will be made to relate such large earthquakes to the patterns of nuclear testing. Such testing is conducted by the United States (USA), the Soviet Union (USSR), France, the United Kingdom (UK) and China. Emphasis was placed on earthquakes of M>=6 because these are the ones that cause considerable property damage and/or kill hundreds of people in short periods of time. Further, the data was more manageable when such magnitude earthquakes were considered. For example, there are between 5,000 to 7,000 earthquakes of M>=4.5 each year around the world, whereas in any given year since 1900, the highest number of earthquakes M>=6 (in 1957) was 214. As the magnitude threshold is lowered, many thousands more small events must be screened. Earthquakes have always been part of the Earth's geologic history. On the other hand, nuclear testing only began in earnest in 1951. In 1963, such testing was moved underground. The greatest recorded earthquake death toll of 830,000 was in Shaanxi, China, in 1556. The worst in this century was on 28 July 1976, when the northeastern Chinese city of Tangshan was levelled and about 000,000 people were killed. That quake measured M 7.8. Coincidentally, five days before the quake (23 July) the French detonated a nuclear bomb in the South Pacific Mururoa Atoll, and one day before (27 July) the USA detonated a nuclear bomb of 20-150 kilotons (KT) at the Nevada test site.

The nuclear era began on 16 July 1945, when Trinity was dropped 100' from a tower near Alamogordo, New Mexico. The yield was 19 KT of TNT equivalent. Soon after this test, on the 5th and 9th of August, the 15 KT nuclear device "Little Boy" was dropped on Hiroshima, and the 21 KT "Fat Man" was dropped on Nagasaki, ending World War II. Since 1945, the major powers have exploded a total of over 1,800 nuclear bombs (through March 1989). An average of close to 50 underground nuclear tests have taken place each year since 1980. There is little doubt that planet Earth is under severe environmental stress. It is not getting any better. Recently the prestigious environmental research group, the Worldwatch Institute, issued their latest "State of the World Report", which shows that the world is being pushed to the brink. "We are losing at this point, clearly losing the battle to save the planet," said the report's chief author, Lester Brown. The impending result, he warned, "will shake the world to its foundation." Ozone depletion, toxic wastes, acid rain, water scarcity and pollution, forest destruction, and topsoil loss are all part of this impending environmental disaster. Perhaps it is high time to consider underground nuclear testing as a part of this infamous list.

Patterns of Earthquakes M>=6, 1900 to 1988

For means of comparing patterns and trends of M>=6 earthquakes with nuclear testing, 1950 will be used as the watershed year. There were no nuclear tests in that year and only nine covering the years 1945 through 1949. The idea is to identify patterns in the first half of this century (1900 to 1949) and compare these to the second half of the century (1950 to 1988). The most evident trend from the graphs in Figure 1 is the change in the comparative number of earthquakes of various magnitudes for the period before and after 1950. The first 50 years of this century recorded 3,419 such earthquakes of M>=6, an average of 68 per year. The last 39 years of this century recorded 4,963 earthquakes of M>=6, an average of 127 per year. In other words, the average per year for such earthquakes has about doubled in the second half of this century as compared to the first half of the century. Also, from 1900 through 1949, there were only 8 years in which there were over 100 earthquakes of M>=6. This entire

* Professor of Geography, University of New Brunswick, Frederickton, New Brunswick, Canada.

Earthquakes & Nuclear Testing   Whiteford   11

" The average number (of earthquakes) in the (M 6.0 to 6.5) range has tripled since 1950 ...."

cluster of 8 was found between 1931 and 1941. The highest number was 182 in 1934, and this compared to a low of 17 in 1904. Starting in 1950, the trend was completely reversed. In this 39 year period, from 1950 to 1988, the overwhelming majority of years had a total of over 100 earthquakes of M>=6. Again, this compares to only 8 years for the first 50 years of this century. The highest number was 214 in 1957, while the lowest was 78 in 1962. It is interesting to note that the years 1959 and 1960 were relatively free of nuclear tests. Coincidentally, two years later, the number of earthquakes M>=6 dropped to only 78 in 1962 and 83 in 1963. These earthquake totals were the lowest for any given year covering the entire second half of this century. When the M>=6 earthquakes are divided into groups, another trend becomes evident.

From 1900 through 1949, there were a total of 101 earthquakes of M>=8, with a yearly maximum of 7 in 1906. Those 1906 earthquakes included the famous San Francisco earthquake of 18-19 April, at M 8.3, which killed over 400 people. But from 1950 through 1988, a total of only 30 earthquakes of M>=8 were recorded. The most in any given year since 1949 were 4 back in 1950. This included the great Indian earthquake on 15 August at M 8.7, which killed over 1500 people. Thus, for the first half of this century 101 earthquakes of M>=8 were recorded, as compared to only 30 of such earthquakes for the second half of the century. And for the last 10 years, there have been only three earthquakes M>=8 recorded. The last was on 20 October 1986, when a M 8.3 earthquake struck the Kermadic Islands of the South Pacific. This quake happened just 4 days after the USA exploded a 20-150 KT bomb in Nevada, on 16 October.

It appears, therefore, that given such an increase in earthquakes of M>=6 since 1950, and a decrease in earthquakes of M>=8, that the observed increase must have occurred in between the two magnitude ranges. In fact, the increase has mostly occurred in the M 6.0 to 6.5 range, as seen in the table below. The average number in this range per year has tripled since 1950, from 24 to 72, as compared to earthquakes between M6.5 to <7.0. The numbers for earthquakes M>=7 have dropped since 1950, relative to the first half of the century. There were 1145 (average of 22 per year) from 1900 to 1949, and only 699 (average of 17 per year) from 1950 to 1988.

It should be noted that the ability to locate earthquakes in the world has increased dramatically since the turn of the century because of improved global communications and seismograph instrumentation. A dramatic increase in the number of recording stations has also occurred. For example, about 350 seismograph stations were operating in 1931, whereas today there are over 3,000 active stations around the world. It is generally conceded, however, that the largest earthquakes have been recorded relatively consistently since 1900, and these factors could have only a small effect on the number of events located per year for magnitudes above 6.0.

In conclusion, since 1950 the trend of the earthquakes M>=6 is as follows: There have been 1500 more in the last half of this century compared to the first half, and the average per year has doubled. Further, the increase has been most dramatic in the M 6.0 to 6.5 range, while a dramatic drop is seen in earthquakes of M>7.0. The question remains as to whether this trend will continue.

Patterns of Nuclear Testing, 1945 to 1989

The table below compiles the respective summary totals of nuclear explosions by country since 1945, while Figure 2 breaks these numbers down by year and type of explosion. The USA and USSR account for about 87% of the total.

Total Nuclear Tests by Country, 1945 to 1988						
Years	USA	USSR	Fra.	UK	Chin	India
1945 - 1962:	304	166	6	23	0	0
1963 - 1988:	629	451	165	18	32	1
Totals:	933	617	171	41	32	1
Grand Total: 1795						

Nuclear Testing began in earnest in 1951 when the USA exploded 16 bombs. They later tested 77 times in 1958, half in the South Pacific and about half at the Nevada test site. In 1962, a record of 98 USA bomb tests occurred, including a 600 KT bomb from a Polaris A2 rocket in the South Pacific. The largest nuclear test explosion conducted by the USA was a 15 megaton (MT) bomb detonated at Bikini Atoll, Marshall Islands in the South Pacific, on 28 February 1954. The largest nuclear test by any country is believed to be a 58 MT bomb detonated by the USSR on 30 October 1961, above the high Arctic island of Novaya Zemlya. Since 9 November 1962, all USA nuclear tests have been conducted underground at the Nevada test site. In 1962, a large number of nuclear tests were carried out (98 by the USA, 44 by the

Earthquakes of Magnitude 6 or Greater				
Magnitude	1900-1949	Average	1950-1988	Average
6.0 to <6.5	1164	24	2844	72
6.5 to <7.0	1110	22	1465	37
>7.0	1145	23	699	18
>8.0	101	2	30	<1

Number of Earthquakes of Magnitude 6 or Greater

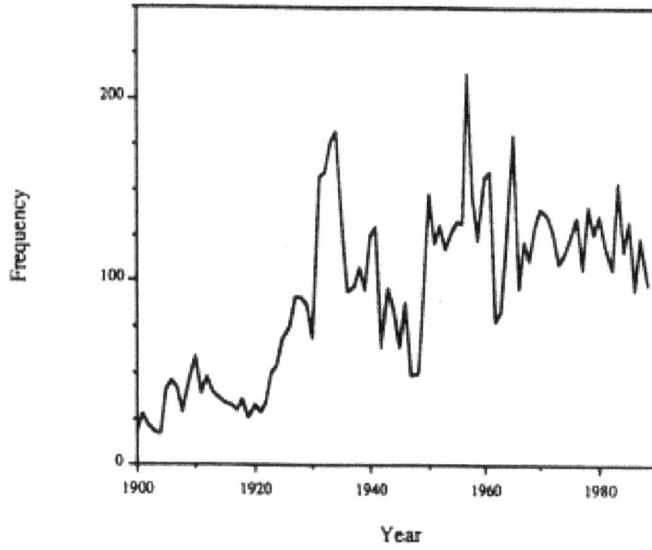

Number of Earthquakes of Magnitude 6.0 to < 6.5

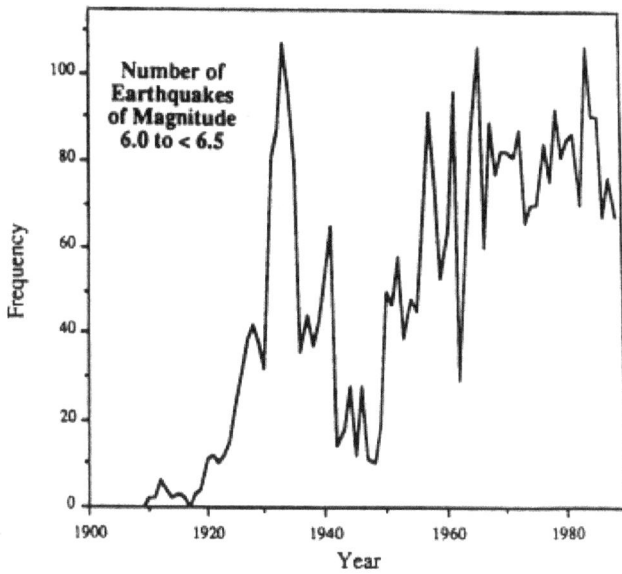

Number of Earthquakes of Magnitude 6.5 to < 7

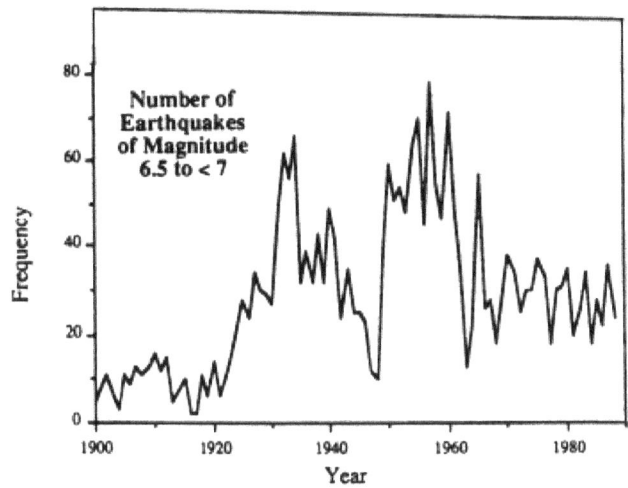

Number of Earthquakes of Magnitude 7.0+

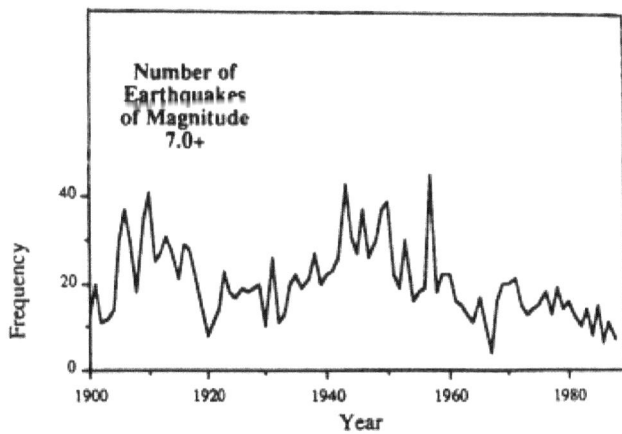

Number of Earthquakes of Magnitude >8

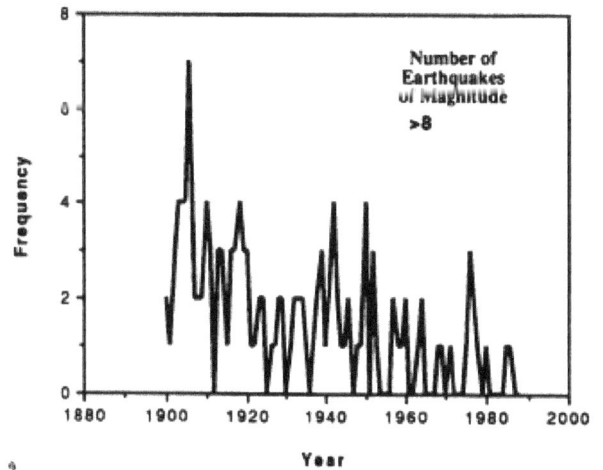

USSR) in anticipation of a halt to above-ground testing, which was a result of the Limited Test Ban Treaty signed in 1963. The French, however, continued to test above the water at the Mururoa Atoll until 1975. And the Chinese did likewise, testing some 16 times above ground at the Lop Nor test site in Sinkiang Province, until 1975. Tests are now limited to a maximum yield of 150 KT, under terms of the Threshold Test Ban Treaty signed by President Richard M. Nixon and Soviet Premier Leonid Brezhnev in Moscow, on 3 July 1974. The ban did not take effect until 31 March 1976, and remains unratified by the US Senate. Testing was stopped completely in 1959 and 1960, and the USSR unilaterally stopped their testing, during a self-imposed moratorium, for 19 months between July 1985 to February 1987. During that time, the USA conducted 26 nuclear tests. Since 1963, nuclear test sites by the five major powers have essentially been confined to the following locations:

Nation	Site Description	Latitude, Longitude	
USA & UK:	Nevada Test Site (65 miles NW of Las Vegas)	37 N	116 W
France:	Mururoa, Fangataufa Atoll (720 miles SE of Tahiti, in Tuamotu archipelago)	22 S	139 W
China:	Lop Nor, Sinkiang Province	41 N	88 E
USSR:	1. Semipalatinsk, Kazakhistan	49 N	78 E
	2. Novaya Zemlya Island Arctic Ocean	73 N	55 E
	3. Ural Mountains, near Serov	60 N	56 E
	4. Siberia, north Lake Baykal	61 N	112 E

The French testing site is very close to the Tropic of Capricorn (23-1/2 S Lat.) and is the only nuclear test site south of the equator. The Soviet Arctic site, the Novaya Zemlya Islands, is the only nuclear test site north of the Arctic circle (66-1/2 N Lat.), and is presently used only once or twice per year. However, from 1958 through 1963, it was the main Soviet nuclear test site. The site was last used on 4 December 1988, when the USSR exploded a nuclear bomb between 20-150 KT. Three days later, on 7 December, the Soviet Armenian earthquake struck, registering M 6.9, and killing upwards of 60,000 people, injuring 13,000, and leaving half a million people homeless. Another of these dangerous coincidences.

The total nuclear tests by all countries since 1945 is 1,795.* The average for the 43-1/2 year period is one test every 8 to 9 days. If the period 1963 through 1988 is taken, the major powers are averaging a nuclear test every 7.3 days. The yearly average in the 1960s was 56; in the 1970s it was 47 tests; and in the 1980s, 47 tests. But the period of July 1985 to February 1987 was the self-imposed test ban by the USSR, so the 1990s should show the yearly level rise to above 50 again, as they try to make up for lost ground. Perhaps the only hope on this nuclear testing path is the attempt to limit nuclear tests to 1 KT, with a view to total elimination. The Soviets emphasized a goal of immediate cessation of all nuclear tests, while the Americans stressed the need to improve verification capabilities, and the need to continue testing in the absence of significant reductions in offensive nuclear weapons. In 1988, each side visited the other side's nuclear test site, to monitor an underground nuclear explosion. The idea was to make sure that both sides can verify whether a test yields more or less than 150 KT. The Soviets prove accuracy by their preferred monitoring method, which counts seismic units such as those used in monitoring earthquakes. In the American method, an electrical cable must be placed within 10 to 15 meters of the blast.

Nuclear Testing and Earthquake Frequencies

The distribution of all earthquakes of M>=5.8 between 1900 through 1949 clearly reflects the boundries of the eleven major tectonic plate zones.** All such earthquakes were located in zones or blocks of 10 degrees of latitude and 10 degrees of longitude, to give frequency and percent distributions on a global scale. The highest percent is the Southern Philippines block at 3.95%, having recorded 135 earthquakes of M>=5.8 during the period. Japan, in the Hokkaido area, is next at 3.07%, with 105 such earthquakes. The table below identifies the major Zones of High Earthquake Frequency, in some cases combining the values for adjacent blocks of latitude and longitude. This convention for aggregating and presenting the data, in percentage values within block of latitude and longitude, will be used throughout the paper.

---

* *This figure will vary somewhat because test dates are gathered by a number of observer agencies, and tests have been confirmed by some, and unconfirmed by others.*

** *Maps of these and other earthquake patterns given in this paper are available from the author.*

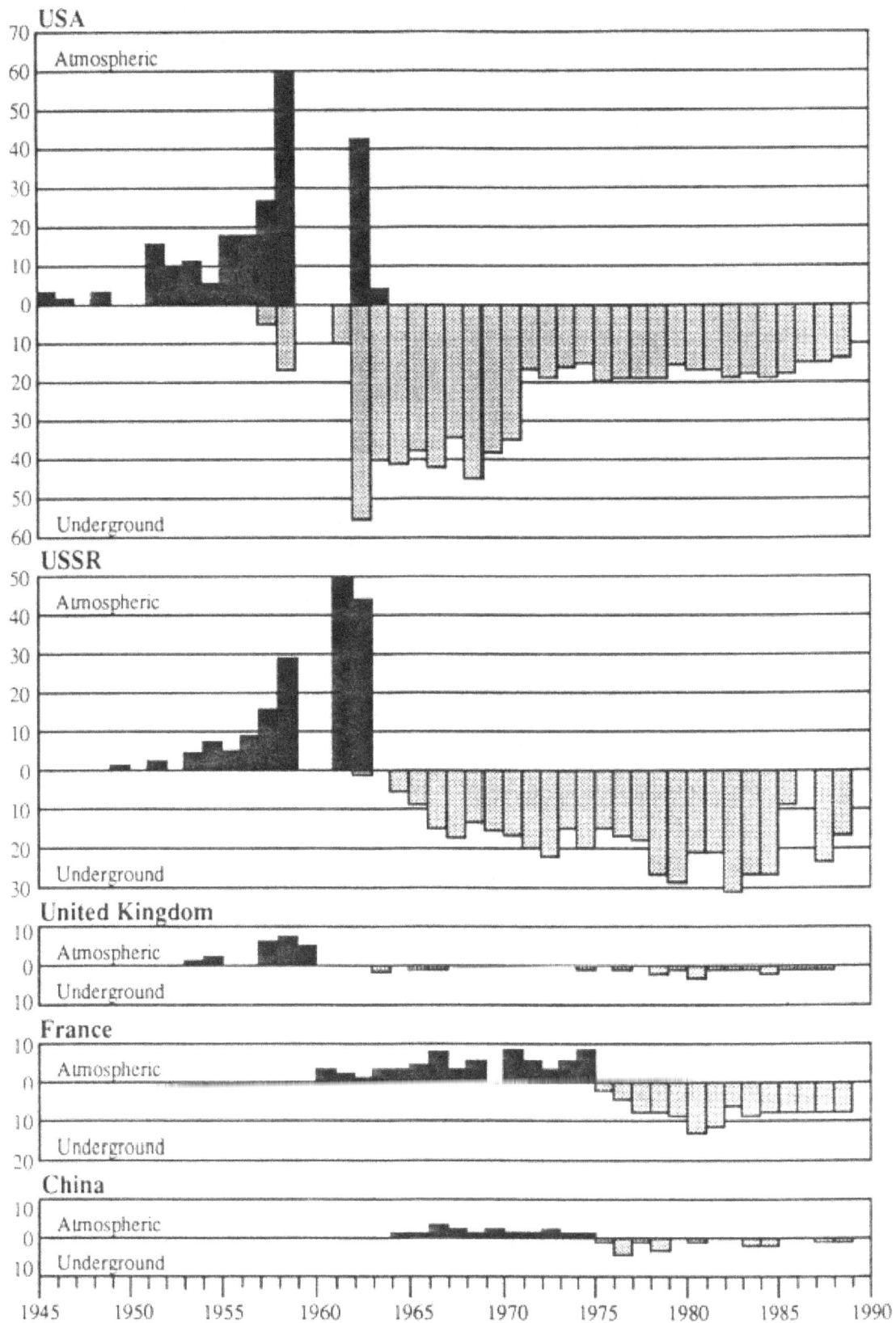

Figure 2.   Atmospheric and Underground Tests from July 16, 1945 to December 31, 1988.

Location	Number of Earth-quakes	Combined no. of Blocks of Lat and Long	% of All Earth-quakes
Indonesia, New Guinea:	424	7	12.39%
Vanuatu, Fiji, Tonga:	377	5	11.03%
Japanese Islands:	310	4	9.07%
Cent. Amer., S. Mexico:	205	4	5.99%
Chile, Peru, N. Argent.:	204	5	5.96%
Philippines:	187	2	5.47%
Taiwan:	74	1	2.16%
N. India, Pakistan:	69	1	2.02%
New Zealand:	43	3	1.25%
S. Greece, W. Turkey:	42	1	1.23%

These area blocks or zones of latitude and longitude received nearly 56% of all the Earth's quakes of M>=5.8 from 1900 to 1949, for a total of 1935 earthquakes. By comparison, all other blocks of latitude and longitude each receive less than 1% of all earthquakes. This established pattern of earthquakes of M>=5.8, with high-lighted zones of high frequency occurrence on a global scale, between 1900 and 1949, will act as a control against which we can compare the patterns of earth-quakes that follow nuclear testing. All such earthquakes were recorded either on the day of the test, or within the four days afterward, for a five day period.

Atmospheric (Above Ground) Nuclear Testing

The first area to consider is the South Pacific. The US tests here totalled 106 and were conducted from 1946 through 1962. The principal sites are listed below:

Location	Number of Tests	Lat	Long
Enewetak	43	11 N	162 E
Christmas Island	24	2 N	169 W
Bikini	23	11 N	165 E
Johnson Island	12	17 N	169 W
Pacific sites	4		

When we review the earthquake data, the following areas emerge as receiving more than their share of the M>=5.8 earthquakes. Hardest hit following USA South Pacific Tests, relative to the pattern prior to 1950, were the blocks of latitude and longitude encompassing Sakhalin Island, the Aleutian Islands, Peru, Bolivia, Central America, Western Samoa, Vanuatu, Baja, California, Hawaii, and Japan.

French testing in the South Pacific covered years 1966 through 1974 when they conducted 44 atmospheric tests, 39 over Mururoa Atoll and 5 over Fangataufa Atoll. The earthquake pattern after the French South Pacific testing is somewhat different, and the zones of high frequency earthquakes following these French tests were Western Samoa, Fiji, the Solomon Islands, the Alaska Panhandle, and the area between the Kamchatka Penin-sula and Aleutian Islands. Interestingly, these regions of high earthquake activity following South Pacific nuclear testing are all confined to the Pacific Ring of Fire, a zone of earthquakes and volcanoes that circles the entire Pacific Ocean. These data are summarized below:

Percent of All Earthquakes M>=5.8
After American and French Above-Ground Nuclear Tests
in the South Pacific, 1946-1975
(Five Day Period)

Location	Pre-Test Period Percent 1900-1949	After USA Tests Percent 1946-1963	After French Tests Percent 1966-1975
W. Samoa:	2.78	4.91	12.50
Fiji:	2.31	2.45	4.17
Kermadec Is.:	0.50	1.84	0
Vanu-Coral Sea:	3.36	6.75	4.17
Solomon Is.:	2.98	1.84	16.67
Hawaii:	0.06	1.23	0
	0.09	1.84	0
N. Peru:	1.02	0.61	4.17
Lima, Peru:	1.26	3.07	0
Bolivia:	0.79	3.07	0
C. Rica, Panama:	1.05	4.91	0
South Mexico:	2.22	4.29	0
Baja, CA:	0.35	2.45	0
Alaska Panhan.:	0.50	0	8.33
W. Alaska:	0.41	1.23	4.17
Aleutian Is.:	0.88	7.36	0
	0.97	3.07	4.17
	0.88	2.45	4.17
	0.76	0.61	4.17
Sakhalin:	0.97	7.98	0
Hokkaido:	3.07	1.84	4.17
Tokyo:	2.46	3.68	0
S.Japan, Bonin:	0.53	2.45	0
Burma:	0.94	0	4.17
Tehran:	0.47	0	4.17

British tests in the South Pacific covered the period 1957 to 1958 and involved only nine tests at Christmas Island (1.7 N Lat, 157 W Long). Twelve British tests were conducted in Australia from 1952 to 1957. Since 1962, all

British tests have been underground at the USA Nevada test site. Only the Christmas Island tests were examined here, but once again, familiar names emerge:

Percent of All Earthquakes M>=5.8
After British Above-Ground Nuclear Tests
in the South Pacific, 1952-1957
(Five Day Period)

Location	1900-1949	After British Tests
Central America &		
S. Mexico:	4.64	20.00
Kermadec Is./Samoa:	7.13	15.00
Solomon Is./Vanuatu:	6.34	10.00

From 1951 until 1963, the USA tested 100 times above ground at the Nevada test site. When these test dates are matched with earthquakes of M>=5.8 within a five day period, the following areas show a higher than normal share of such earthquakes, as compared to the 1900-1950 period:

Percent of All Earthquakes M>=5.8
After USA Above-Ground Nuclear Tests
in Nevada, 1951-1963
(Five Day Period)

Location	1900-1949	After USA Tests
Kermadec/W. Samoa &		
Vanuatu:	11.61	16.05
Kamchatka/Aleutian Is.:	4.46	10.82
Taiwan:	2.16	5.60
Solomon Is.:	2.98	5.22

The Soviet above ground testing involved 166 tests from 1949 to 1962. About 70% of the known Soviet tests have occurred at their two main sites near Semipalatinsk in eastern Kazakhstan (50%), and on the island of Novaya Zemlya, north of the Arctic Circle (20%). The above ground testing was very active at Novaya Zemlya in the years 1958 to 1962. When the 79 tests at that site are matched to the M>=5.8 earthquakes over the five day period for those years, the following areas emerge as high risk zones:

Percent of All Earthquakes M>=5.8
After USSR Above-Ground Nuclear Tests
in Novaya Zemlya, 1958-1963
(Five Day Period)

Location	1900-1949	After USSR Tests
Costa Rica/Panama:	1.05	8.82
Mindanao, Philippines:	3.95	6.86
Vanuatu:	3.36	6.86
Kermadec Is.:	2.81	6.86
Kamchatka:	0.97	5.88
Java/Java Trench:	0.94	8.82
S. Aleutian Is.:	1.85	10.76

These seven areas were struck 55% of the time during a five day period following an above ground test at Novaya Zemlya, for the period 1958 through 1962.

The USSR main testing site today is Semipalatinsk. But from 1945 through 1962, only a total of 53 above ground tests were recorded. This number also included tests conducted at their other mainland sites in Siberia and the Ural Mountains. Two regions are most noticeably tied to this particular Soviet nuclear test site: Taiwan at 9.48%, and the South Aleutian Islands at 23.28%. These two locations respectively received only 2.16% and 0.88% of all M>=5.8 earthquakes during the pre-testing period of 1900-1949. Two other high percent zones are the southern Mexico, Central American Coastal region, at 8.62%, and the western Samoa- Tonga region, at 9.48%. Another very interesting observation was that these mainland Soviet tests appeared to have little significant affect upon earthquake patterns in the Indonesia, Solomon Island region, a zone that appears tied to other above ground nuclear test sites.

Underground Nuclear Testing

Underground, or below ground nuclear testing for the USA and USSR started in 1963. Through 1988 the USA tested underground 629 times, almost exclusively at the Nevada test site. The Soviets tested underground 451 times, mainly at the Semipalatinsk location, but with 35 underground tests at Novaya Zemlya.

**Percent of All Earthquakes M>=5.8
After USA Below-Ground Nuclear Tests
in Nevada, 1963-1988
(Five Day Period)**

Location	1900-1949	After USA Tests
Vanuatu:	3.36	6.81
Solomon Islands:	2.98	4.89
Nevada:	0.61	4.36
Aleutian Is. Block:	3.31	12.57
Santiago, Chile:	0.76	2.79

Observations from the table above show that around 30% of the time, during a given five-day period following an underground nuclear test in Nevada, a M>=5.8 earthquake has hit the Vanuatu Islands, Solomon Islands, Nevada itself, parts of the Aleutians, or Santiago, Chile. The 35 tests at Novaya Zemlya 1963 through 1988 tie to familiar areas again, as given in the table below.

**Percent of All Earthquakes M>=5.8
After USSR Below-Ground Nuclear Test
in Novaya Zemlya, 1963-1988
(Five Day Period)**

Location	1900-1949	After USSR Tests
S. Aleutians:	0.76	9.09
Mindanao:	3.95	9.09
Vanuatu:	3.36	6.06
Hokkaido:	3.07	6.06
Novaya Zemlya:	0	18.18
W. Samoa/Kermadec Is.:	5.59	15.15

The major USSR nuclear test site in Semipalatinsk, and associated mainland sites, has received over 400 underground nuclear tests since 1963. Interesting patterns of earthquakes again follow nuclear tests at these sites. The South Pacific zone once again shows a relationship, but so too does the Nevada region, which previously only appeared with a high percentage following American nuclear tests in Nevada itself. The Aleutian Islands usually show an increase in earthquake frequency following nuclear tests, but this is not the case following a USSR underground test at Semipalatinsk, a marked contrast with the Novaya Zemlya test site. Further, the Semipalatinsk region itself received 3.15% of all M>=5.8 earthquakes following nuclear tests at that site, as compared to a low 0.23% prior to 1950.

**Percent of All Earthquakes M>=5.8
After USSR Below-Ground Nuclear Tests
in Semipalatinsk and Other
USSR Mainland Sites 1963-1988
(Five Day Period)**

Location	1900-1949	After USSR Tests
Solomon Is.:	2.98	8.92
Vanuatu:	3.36	6.49
Fiji, New Caledonia, Kermadec,W.Samoa block:	8.33	14.86
Honshu-Kyushu:	2.19	2.43
Nevada:	0.61	1.62
Aleutians:	2.55	3.24
Semipalatinsk:	0.23	3.51

The French have conducted 111 underground nuclear tests at the Mururoa Atoll in the South Pacific, from 1975 through 1988. As pointed out by a recent National Resources Defense Council paper, the French have accounted for some 20% of all nuclear tests within the last 10 years. Serious fractures of the coral atoll and constant nuclear contamination of the site and surrounding waters has occurred. Some observers feel that the French will have to move these tests to the nearby Fangataufa Islands. When the earthquakes of M>=5.8 are mapped within 5 days following all French underground testing, many Pacific regions again emerged as most affected and noticeable.

**Percent of All Earthquakes M>=5.8
After French Below-Ground Nuclear Tests
at Mururoa Atoll, South Pacific 1975-1988
(Five Day Period)**

Location	1900-1949	After French Tests
W. Samoa, Kermadec Is.:	5.79	12.41
New Britain, Solomon Is.:	2.98	10.08
S. Aleutian zone:	2.61	7.76
Vanuatu:	3.36	9.30
Hokkaido:	3.07	5.43
Taiwan:	2.16	3.10
Mexico City, El Salvador:	2.22	3.10
Columbia:	0.85	2.33

Interestingly, the French nuclear test site itself at Mururoa Atoll (22 S Lat, 139 W Long.) has a low 1.55% earthquakes observed in the five day period following an underground test. This is quite different when compared to earthquake frequencies following underground tests at the three other test sites in Nevada (4.36%), Semipalatinsk (3.51%), and Novaya Zemlya (18.18%). This might be attributed to the fact that energy released from an underground nuclear test dissipates differently from a test site surrounded by an ocean.

Atmospheric and Underground Nuclear Tests, Combined

When all the above ground (atmospheric) nuclear explosions are considered, certain areas of the world reveal high frequency patterns of the M>=5.8 earthquakes following such tests. These areas, listed below, account for about 50% of all the M>=5.8 earthquakes that followed an above ground nuclear explosion within the five day period.

Percent of All Earthquakes M>=5.8 After All Atmospheric, Above-Ground Nuclear Tests, Worldwide, 1950-1974 (Five Day Period)

Location	1900-1949	After Nuclear Tests
Kamchatka, Aleutians:	0.88	2.06
	0.97	2.06
	0.76	1.65
	0.88	7.27
	0.97	4.66
Vanuatu:	3.36	6.86
Solomon Is.:	2.98	4.39
Panama, Costa Rica:	1.05	4.12
S. Mexico, El Salvador:	2.22	3.57
Kermadec Is.:	0.50	3.43
	0.03	1.23
Taiwan:	2.16	3.29
S. Greece, Turkey:	1.23	2.06

The earthquake patterns following all underground nuclear tests is similar to that for the above ground testing, and when grouped together, account for about 50% of all the M>=5.8 earthquakes that follow an underground nuclear explosion. It is interesting to note that the Nevada area has a 2.65% chance of having such an earthquake. This compares to 0.61 percent for the pre-nuclear 1900-1949 period, and is slightly more than Taiwan's 2.47%.

Percent of All Earthquakes M>=5.8 After All Underground Nuclear Tests, Worldwide, 1963-1988 (Five Day Period)

Location	1900-1949	After Nuclear Tests
Kamchatka, Aleutians:	0.88	1.06
	0.97	0.97
	0.76	3.35
	0.88	2.12
	0.97	1.85
	0.70	1.76
Vanuatu:	3.36	6.70
Solomon Is.:	2.98	6.70
Fiji, Tonga, Kermadec Is.:	2.05	1.59
	2.78	3.88
	2.31	3.88
	0.50	1.68
Hokkaido:	3.07	3.26
Papua, New Guinea:	2.22	3.09
Nevada:	0.61	2.65
Taiwan:	2.16	2.47

The continuation of underground nuclear bomb testing mainly by the USA, USSR, and France, should alert certain areas of the world to note when and where the tests occur. The table below summarizes the patterns of earthquakes of M>=5.8 within a five day period, following nuclear tests at various test sites.

Percent of Earthquakes M>=5.8 After Nuclear Tests at Various Test Sites (Five Day Period)

Area Affected	After USA /Nevada test	After USSR Semipala. test	After USSR Zemlya test	After French Mururoa test
Aleutian Is.:	12.57	3.24	9.09	7.76
Samoa, Kermadec:	----	14.86	15.15	12.41
Vanuatu:	6.81	6.49	6.06	9.30
Solomons, N. Brit:	4.89	8.92	----	10.08
Mindanao, Philip:	2.09	----	9.09	----
Taiwan:	2.09	---	----	3.10
Papua, New Guin:	2.79	----	----	----
Hokkaido:	3.66	----	6.06	5.43
Honshu-Kyushu:	----	2.43	----	----
Novaya Zemlya:	----	----	18.18	----
Semipalatinsk:	----	3.51	----	----
Nevada:	4.36	1.62	----	----
Mexico, El Salv:	----	----	----	3.10
Santiago, Chile:	2.79	----	----	----
Totals:	**42.05%**	**41.07%**	**63.63%**	**51.18%**

Two areas show a significant tie to nuclear testing, regardless of who tests, notably the Aleutian Islands chain, and the South Pacific area inclusive of Vanuatu, the Solomon Islands, Western Samoa, and the Kermadec Islands. Further, the Japanese islands of Hokkaido, Honshu and Kyushu should note who tests and when. The island of Mindanao should take some precautions when the Soviets test at Novaya Zemlya. For Nevada, it is necessary not only for them to monitor American tests, but also Soviet Semipalatinsk tests, including the Urals and Siberia sites. The two most vulnerable areas in the South Pacific, the Solomon Islands and Vanuatu, should monitor tests weekly. Each of these areas have up to a 10% chance per week of having a M>=5.8 earthquake, because nuclear tests in the 1980s are conducted on an average of one per week.

The "Killer Earthquake" and Nuclear Tests

Of all the earthquakes that do occur, the most frightening of them is the one identified as the "killer quake". It can be defined as an earthquake which kills at least 1,000 people. It is especially interesting to note a dangerous coincidence when all the killer earthquakes since 1951 are simply listed and matched to the dates of nuclear tests. The table below displays the "match" between nuclear explosions and killer earthquakes.

Killer Earthquakes, 1951 - 1988, Matched with Nuclear Tests (Five Day Period)

# Tests Per Year	Nuclear Test Date	Earthquake Date	Location	Magnitude	Deaths	Test/Quake Match?
17	1953: Mar.17	Mar.18	NW Anatolia	7.2	1,200	yes
33	1956: Jun.6-16 (5 separate tests)	Jun.10-17	Kabul, Afghanistan	7.7	2,000	yes
54	1957: — —	Jul. 2	Iran	7.4	2,500	no
	1957: Dec.9	Dec.13	Iran	7.2	2,000	yes
3	1960: — —	Feb.29	Agadir, Morocco	5.8	12,000	no
	1960: — —	May.22	Arauco, Chile	>8.3	5,000	no
145	1962: Sep.1	Sep.1	Buyin-Zara, Iran	7.1	13,000	yes
47	1963: — —	Jul.26	Skoplje, Yugoslavia	6.0	1,100	no
67	1966: Aug.19	Aug.19	Varto, Turkey	6.9	2,600	yes
64	1968: Aug.27,29	Aug.31	Dasht-E Bayaz, Iran	7.4	12,000	yes
61	1970: Mar.26,27	Mar.28	Gediz, Turkey	7.4	1,100	yes
	1970: May 28,30	May 31	Chimbote, Peru	7.7	68,000	yes
46	1972: Apr.11??	Apr.10	Iran	6.9	5,100	??
	1972: Dec.21	Dec.23	Managua, Nicaragua	6.2	5,000	yes
46	1974: Dec.27	Dec.28	Pattan, Pakistan	6.3	5,200	yes
38	1975: Sep.6	Sep.6	Lice, Turkey	6.8	2,300	yes
45	1976: Feb.4 (2)	Feb.4	Guatemala City	7.5	23,000	yes
	1976: — —	May 6	Italy	6.5	1,000	no
	1976: Jul.27	Jul.28	Tangshan, China	8.2	800,000	yes
	1976: — —	Aug.17	Mindanao, Philip.	7.8	8,000	no
	1976: Nov.23 (2)	Nov.24	Eastern Turkey	7.9	4,000	yes
46	1977: — —	Mar.4	Bucharest, Romania	7.5	1,600	no
59	1978: Sep.13,15	Sep.16	Tabas, Iran	7.7	25,000	yes
55	1979: — —	Dec.12	Colombia-Ecuador	7.9	800	no
55	1980: Oct.8	Oct.10	Al Asnam, Algeria	7.3	4,500	yes
	1980: — —	Nov.23	Naples, Italy	7.2	4,800	no
57	1982: Dec.10	Dec.13	Dhamar, N. Yemen	6.0	2,800	yes
57	1983: Oct.26	Oct.30	Posinier, Turkey	7.1	1,300	yes
35	1985: — —	Sep.19	Mexico City	7.9	10,000	no
24	1986: — —	Oct.10	El Salvador	5.4	1,000	no
40	1988: Nov.5	Nov.6	Burma, China	7.3	1,000	yes
	1988: Dec.4	Dec.7	Armenia, USSR	6.8	60,000	yes

Each of the 32 killer earthquakes which struck between 1951 and 1988 caused at least 1,000 deaths, with the worst being 800,000 killed in the 1976 magnitude 8.2 China earthquake. This China earthquake was the worst for deaths recorded in this century, and coincidentally the USA tested a nuclear bomb one day before the earthquake hit. Over the 37 years of nuclear testing, 20 of the 32 killer earthquakes, or 62.5%, occurred on the same day, or within four days of a nuclear test. The total death toll for these 20 killer earthquakes is over 1 million people. The following table shows the breakdown of these 20 killer earthquakes.

**Twenty Killer Earthquakes
Matched with Nuclear Tests, 1951-1988**

Number of Quakes	Days after Nuclear Test
12	Same day, or 1 day later
3	2 days later
2	3 days later
3	4 days later

(When two or more nuclear tests occur prior to a killer quake, only the test closest to the quake date is counted.)

Is this pure coincidence?

Conclusions

Some people would question the idea of directly linking nuclear testing with the pattern of large, powerful earthquakes which follow within days of a test. Given the large number of such earthquakes per year, and the high number of nuclear tests per year, there might be a chance match between any given test and the occurrence of a large earthquake. In the 1980s, there were an average of 47 tests and 120 earthquakes of M>=6, per year. While a chance correlation might appear to be at work, the geographical patterns in the data, with a clustering of earthquakes in specific regions matched to specific test dates and sites, do not support the easy and comforting explanation of "pure coincidence". The phenomenon clearly requires further study. The primary purpose here was to identify frequency patterns of earthquakes following a given nuclear test. One does not expect nuclear testing to stop. However, what is needed is a full disclosure of when and where tests will take place. This way, certain key areas of the world can ready themselves for the possible M>=5.8 earthquakes. It is imperative that the press become much more vigilant in alerting the world to nuclear bomb tests. If these tests are occurring on average once every week, the public has a right to be informed. And obviously, more study and research of the question of a link between nuclear bomb tests and earthquakes is needed. This effort is simply a beginning.

Selected References on the Environmental Effects of Nuclear Explosions:

* Babst, D.V. & Dely, A.: "Nuclear Testing and Volcanic Activity", *Journal of Peace Research*, Brandon, Manitoba, June 1987, p.63-69.
* Bolt, B.A.: *Nuclear Explosions and Earthquakes, the Parted Veil*, W. H. Freeman, San Francisco, 1976.
* Dahlman, O., et al: "Ground Motion and Atmospheric Pressure Waves from Nuclear Explosions in the Polynesian Test Area Recorded in Sweden, 1970", FOA 4 Report, C4461-26, 1971.
* Danielson, B. & Danielson, M.: *Poisoned Reign*, NY, Penguin Books, 1986.
* Davidson, C.I., et al: "Radioactive Cesium from the Chernobyl Accident in the Greenland Ice Sheet", *Science* 237:633-634, 7 August 1987.
* Fieldhouse, R.W., et al.: "Nuclear Explosions", *World Armaments and Disarmament*, SIPRI Yearbook, 1987, New York, Oxford U. Press, 1987, p.45-52.
* Fuller, J.G.: *The Day We Bombed Utah*, NY, New American Library, 1984.
* Goldblat, J. & Cox, D.: *The Debate About Nuclear Weapon Tests*, Ottawa: Canadian Institute for International Peace and Security, 1988, Occasional Paper #5.
* Harris, D.L.: "Effects of Atomic Explosions on the Frequency of Tornadoes in the US", *Monthly Weather Review*, Dec. 1954, pp.360-369.
* Kato, Y.: "Recent Abnormal Phenomena on Earth and Atomic Power Tests", *Pulse of the Planet*, 1:5-9, 1989.
* "Lightning Increase After Chernobyl", *Science News*, 312:238, 10 October 1987.
* Lall, R.G. and Brandes, P.D.: *Banning Nuclear Tests*, NY, Council on Economic Priorities, 1987.
* Malcolmson, R.W.: *Nuclear Fallacies, How We Have Been Misguided Since Hiroshima*, Kingston, Ontario: McGill-Queen's University Press, 1985.
* McEwan, A.C.: "Environmental Effects of Underground Nuclear Explosions", *World Armaments and Disarmaments, SIPRI Yearbook 1988*, NY, Oxford U. Press, 1988, p.75-91.
* O'Keefe, B.J.: *Nuclear Hostages*, Boston, Houghton Mifflin, 1983.
* Reich, W.: *The Oranur Experiment, First Report (1947-1951)*, Wilhelm Reich Foundation, Maine, 1951 (partly reprinted in Reich, W.: *Selected Writings*, Farrar, Straus & Giroux, 1960)
* Smith, R.J.: "Scientists Implicated in Atom Test Deception", *Science* 218:545-547, 5 Nov. 1982; "Atom Tests Leave Infamous Legacy", *Science*, 218:266-269, 15 Oct. 1982.
* Titus, A.C.: *Bombs in the Backyard*, Las Vegas, U. of Nevada Press, 1986.
* Trombley, A.: "An Interview..." *Wildfire*, p.17-33, January 1988.
* Matsushita, S., et al: "On the Geomagnetic Effect of the Starfish High Altitude Nuclear Explosion", *J. Geophysical Res.* 69:917-945, 1964.

The Psycho-Physiological Effects of the Reich Orgone Accumulator

Stefan Müschenich, D. Psych. & Rainer Gebauer, D. Psych.

Abstract of a Diploma Thesis, University of Marburg, West Germany, 1986. Copyright (C) 1986, All Rights Reserved by Stefan Müschenich and Rainer Gebauer, Marburg, West Germany.

Editor's Note: Since the acceptance of this Thesis by the Faculty of Psychology at the University of Marburg in 1986, Muschenich and Gebauer have published a German-language version of the work, titled Der Reichsche Orgonakkumulator: Naturwissenschaftliche Diskussion, Praktische Anwendung, Experimentelle Untersuchung (Nexus Press, Fichardstr. 38, 6000 Frankfurt 1, West Germany,1987; Available in the USA from Natural Energy Works, PO Box 864, El Cerrito, CA 94530). They have additionally presented their work at various professional meetings and workshops in Europe. Their pioneering study appears to have been the first in academic circles to concern itself with the orgone energy accumulator, and they are to be congratulated for a job well done. Since the publication of their work, we have become aware of other research studies on the human physiological response to the accumulator that have been undertaken; one of these was recently published (see the "Orgonomic Research Review" section), and others will be out within the next year.

The above mentioned Thesis contains an experimental investigation of the psycho-physiological changes in volunteer test subjects during sessions in a Reich orgone accumulator, conducted at the University of Marburg, West Germany.

Proceeding from clinical observations as well as from further research work in the biological and physical fields, the Austrian-born physician and psychotherapist Wilhelm Reich postulated a specific bioenergy manifesting itself in the living organism, and called it orgone. This hypothesis was based on extensive biophysical studies conducted by Reich between 1934 and 1957 at the universities of Oslo and New York, and in his own research laboratories. Steps in this development were experiments on changes in endosomatic skin potential due to certain stimuli, the description of energy-carrying vesicle-like structures (called "bions") during the microscopic observation of disintegrating biological slides, and investigations of hitherto unexplained atmospheric energy phenomena that play a part in forming weather conditions. Finally Reich claimed that the orgone was a universally existing kind of energy and he attributed to it a specific (psycho-) physiological effectiveness on the human organism.

About 1940 Reich published the construction plans for an apparatus that was able to concentrate this energy within its interior. He postulated that spending some time in this "orgone accumulator" (ORAC) produced certain psychic and somatic reactions. Later he used the device for therapeutic purposes in the treatment of some syndromes. An orgone accumulator may be described as a closet-like structure or box, each of its walls consisting of a number of alternate layers of organic material (for the exterior) and metal (for the interior).

In the theoretical part of our Thesis we first discussed a choice of empirical studies published by various scientists who dealt with the physical aspects of the atmosphere within the device. The phenomenon of a constantly increased air temperature in the accumulator (the "To-T" effect), the observation of a delayed electroscopical discharge rate, and processes connected with alterations in air humidity and water evaporation rate are of special interest in this context. Furthermore some medical case histories were demonstrated to illustrate the effects of the orgone accumulator that were conducive to health during the therapy of various diseases. The studies mentioned were described in their topical and historical connection with Reich's research work and concepts. In doing so, we critically discussed the scientific validity of the theories involved.

The main effort of our own experimental work was to investigate the psycho-physiological effects that are attributed to orgone accumulator sessions. Reich claimed that body temperature rose during ORAC sessions and he described a general vagotonic activation due to sitting

The authors may be contacted as follows:
Stefan Müschenich, ~~Fischteich 26, D-3550 Marburg-Gisselberg~~, West Germany; Rainer Gebauer, ~~Steinweg 5, D-3571 Amoneburg~~ / Erfurtshausen, West Germany.

in the device for a certain period of time. Considering the contents of orgonomic publications, and the results of our own pilot tests, we decided to explore the systematic changes in body core temperature, skin temperature, and heart rate (ECG). As far as we know, the two last mentioned parameters have never before been evaluated in a scientific manner.

We conducted a long-term study with 15 volunteer subjects, each of them carrying through 20 experimental hours. In this experiment the physiological variables mentioned above were continuously recorded during the sessions with the help of electronic devices. Ten persons undertook ten 30-minute sessions in an eight-fold coated orgone accumulator built according to Reich's instructions. Besides, they carried through ten 30-minute sessions in an almost identical-looking control box. This dummy had been constructed by us for purposes of comparison. It consisted only of organic matter, but concerning its size, shape, and insulating properties it did not differ from the original box. Five additional subjects conducted all their 20 sessions in the same box: three persons used the orgone accumulator every time, while two subjects used only the control box. Before each session the subjects sat in a comfortable relaxation-chair for 15 minutes. By this provision we wanted to establish a standardized psychophysiological initial level. During this space of time the physiological data mentioned were already measured. First, this allowed a comparison between the physiological reaction patterns in the orgone accumulator and in the control box . Second, one could relate the data recorded during the sessions in one of the two boxes to the activation standard previously measured in the relaxation-seat. The entire investigation was conducted as a "double-blind study", which means that neither the volunteer subjects (who had been chosen at random) nor the persons that gave the instructions and recorded the data knew anything about the experiment they were taking part in. The clothing of the subjects, the position of the two boxes, the sequence in which the two devices were used, and the other experimental modalities were standardized or balanced out. In contrast for example to the medical case histories mentioned above, we took into consideration psychological sources of error, and artifacts caused by superimposition effects. Intending to control these factors, we had the persons fill out a questionnaire we had designed after each session, which revealed information about their psychophysiological sensations and moods during their stay in the boxes. This exploration, as well as the fact that the subjects were absolutely uninformed, was to eliminate falsifications produced by suggestion (e.g. by the researchers undertaking the study) or autosuggestive factors.

In this way we evaluated the subjective quality of potential physiological changes. Additionally, the recorded ECG-data conveyed information about the psychic and emotional excitement of the participating persons. Meteorological and physical variables were also continuously measured. They were to be correlated with the psychophysiological data. The investigation gave the following results:

All of the (a priori formulated) hypotheses, which claimed that there were no greater psychophysiological alterations between relaxation-chair and orgone accumulator than there were between relaxation-seat and control box, could be rejected with a statistical significance on the 1%-level. This means that one may proceed with a probability of 99% on the assumption that, compared with the initial standard during the stay in the relaxation-chair, the physiological data recorded in the orgone accumulator were subject to greater alterations than the data recorded in the control Both the body temperature variables showed a distinct increase during the accumulator sessions. These facts are in accordance with Reich's predictions of a rise in core temperature and an increase in parasympathetic activation produced by orgone accumulator treatment. The interpretation of heart rate pattern, however, is more difficult. While one would expect a decrease, the heartbeat frequency revealed a clear increase between the initial level and the stay in the accumulator. These problems may probably result from the fact that the ECG-data showed a

Figure 1.

Average Change in Body Core Temperatures
The data is for both the control box and the orgone accumulator from among different volunteer subjects participating in the study (identified by number).
Black = Change in body core temperature between relaxation chair and control box session.
White = Change in body core temperature between relaxation chair and orgone accumulator session.

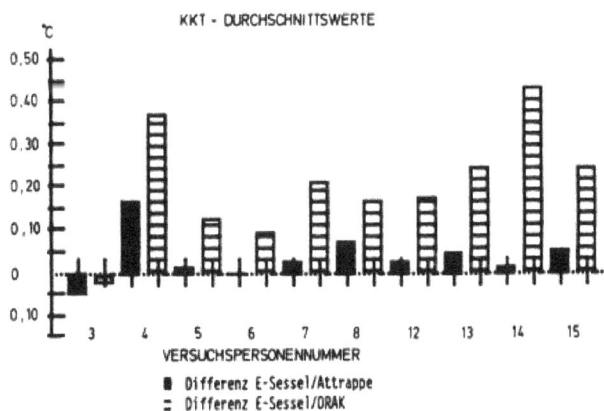

relatively high statistical variance and were much more sen-sitive to accidental external influences and momentary psychic states than the more stable temperature parameters. Further follow-up studies are to explore to what extent a possible ECG-effect may have been superimposed by cognitive processes, nervousness, or anxiety. Still, we can conclude that the assumption of vegetative changes during orgone accumulator sessions was strongly confirmed by our data. The impression one gets from the results of the first 10 subjects was corroborated by the data of those five persons who used only the accumulator or only the control box every time. The number of these last-mentioned cases, however, is too small for a statistical test of significance.

Some further interesting effects are worth mentioning, too: for example, one person seemed to be "resistant" to the orgone accumulator effects, while another proved to respond extremely sensitively to the accumulator. Not only the objective physiological results but the subjective sensations (expressed in the questionnaire) of these two persons justified this hypothesis. Additionally, the "ORAC-resistant" subject was the only one who said that he felt better in the control box than in the orgone accumulator. All other persons preferred the accumulator. Generally it can be said that the questionnaire data corresponded well to the recorded physiological changes. In the accumulator box the volunteer subjects noticed more perceptions of warmth, prickling and tingling on the skin surface; addi-

tionally, they connected more pleasant cognitive associations with the orgone accumulator than with the control box.

In our study a correlation between psychophysiological alterations and meteorological/physical processes was corroborated. During late spring and early summer months the somatic reactions were stronger than during the colder period. Especially the air pressure outside the building seems to be quite a good predictor for the physiological patterns in the accumulator. The phenomenon of a constantly positive To-T difference was statistically confirmed. The total of the meteorological data revealed that the air temperature measured in the orgone accumulator differs from that recorded in the control box on the 1% level.

As a conclusion it can be stated that the results received in our investigation furnish evidence for the assumption that the physical properties of the orgone accumulator and its psychophysiological efficacy on human organisms, postulated by Reich and his associates, factually exist. Various, more natural-scientific oriented follow-up investigations are expected to determine whether the hypothesis is justified that a hitherto unknown biophysical energy (the orgone) is the causative factor of the phenomena described. Additionally, a more extensive study might clarify one or another effect that could not be fully explained by our experiments.

Figure 2.
Changes in Body Core Temperature of Subjects Moving from Relaxation Chair into Control Box. Note the absence of any significant differences in temperatures (except for subject #14).

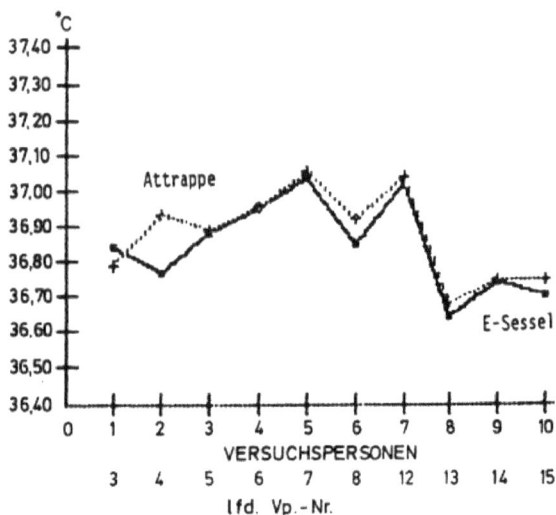

Figure 3.
Changes in Body Core Temperatures of Subjects Moving from Relaxation Chair into Orgone Accumulator. Significant temperature differences are now quite apparent (especially subjects #4 and #14).

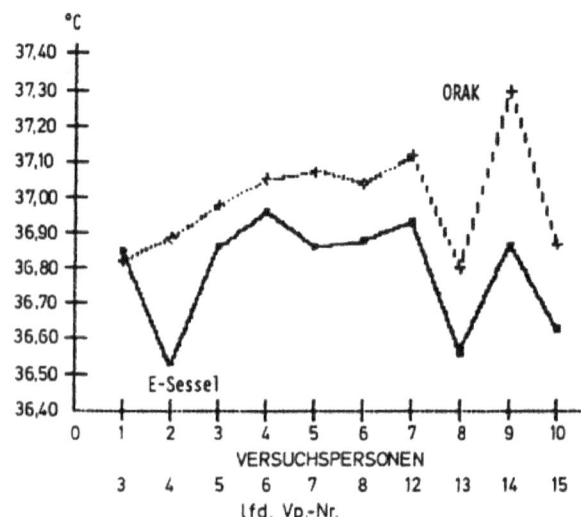

Wilhelm Reich:
Discoverer of Acupuncture Energy?

Prof. Dr. Bernd Senf *

Text of a lecture given at the Third World Congress of Acupuncture, West Berlin. Translated into English by David F. Mayor from a prior publication. Copyright (C) *American Journal of Acupuncture*, 7(2):109-118, April-June 1979.

While the practical therapeutic success of acupuncture in the treatment of psychosomatic diseases is less and less contested, confusion still reigns as to the physical properties of that energy which—according to Chinese teachings—underlies acupuncture. Even today it remains true that acupuncture is frequently dismissed by reference to "suggestion". This makes it all the more important to put the incontrovertible therapeutic successes of acupuncture on a scientific basis. It seems to me that Wilhelm Reich's bioenergetic researches, whose far-reaching consequences have been mostly ignored up to now, are of extraordinary relevance to a scientifically based understanding of acupuncture.

In what follows, I refer briefly to some of Reich's experimental findings that relate to the discovery of a biological energy. I have applied these findings in a series of tests on 150 people, which indicate, in my estimation, the identity of acupuncture energy and the orgone energy discovered by Reich.

On the Origins of Reich's Investigations

Wilhelm Reich (a pupil of Sigmund Freud) started from psychoanalysis and discovered in the course of a growing clinical practice the connections between neurotic (e.g. psychosomatic) diseases and energy blocks in the organism. He observed that, on dissolution of psychic tensions in the course of psychotherapy, muscular tensions were released at the same time. This led him to the hypothesis that muscular tensions are only the physiological indications of psychic tensions. From this he developed a therapeutic method that worked directly with the release of muscular tension (*Vegetotherapy*) and that, as a result, not only demolished muscular *armoring* (that itself is the cause of numerous psychosomatic diseases), but in the same process also dissolved the *character armor*. In the course of this treatment, patients again and again reported that, on release of tensions, they perceived definite *streaming sensations* in the body, though these, it is true, would at first immediately meet with further armor blocks, or new tensions. As the muscular armoring—strongest generally in the diaphragm—was further demolished, streaming sensations would build up throughout the body. In the same process, not only did psychic and psychosomatic disease become less apparent, but at the same time the patient recovered a vitality and spontaneity of movement and sensation in his own body, and in contact with his surroundings. A further concomitant was what Reich called *orgastic potency*, the attendant ability to abandon oneself completely in the sexual act to involuntary muscular movements and intense, pleasurable streaming feelings. So long as the capacity for full orgasm was not established, a fixed quantum of surplus energy would circulate time and again around the organism, leading to the creation of blocks and their corresponding psychosomatic symptoms.

These observations led Reich to the hypothesis that sexual energy is congested through muscular tension, as if dams are being set up which constrict the flow of energy and so lead to ever stronger blocks. His further investigations led him to probe the streaming sensations. In the 1940s, he discovered, in connection with these, an energy form unknown until then and which, in its physical properties, differed essentially from all known energy forms. This—by analog with the word "organism"—he called *orgone energy*. Reich discovered ways of measuring this energy, and persistently found that it showed as an increase in energy potential at the surface of the body in people capable of strongly feeling pleasurable sensations, while the presence of anxiety led to a rapid decrease in potential. Correspondingly, in strongly armored people who, for their part "felt" completely insensible and numb, there registered only a very feeble flicker in the

* Prof. Political Econ., Fachhochschule fur Wirtschaft, ~~Karlsbergallee 25E, D1000 Berlin 22,~~ West Germany.

> " If my hypothesis is correct, ... it must be possible to expose single acupuncture points to concentrated orgone energy ... without needling or even so much as touching the points."

potential. These observations led Reich to interpret emotion as an expression of the orgone streamings in the organism, and ultimately his microbiological investigations led him to the finding that similar fluctuations are already operative in the single-cell organism: in danger-free situations there is an expansion of the cell plasma and of the orgone energy field; in dangerous situations, the plasma and energy field contract.

While Wilhelm Reich started out here, from the viewpoint that all living organisms are endowed with this energy, he later discovered that all space is also filled with orgone energy, though in a different concentration. By making use of the physical properties of this energy that he had discovered, Reich managed at last to accumulate orgone energy from the atmosphere in a controlled manner. The apparatus he constructed to do this he called the *orgone accumulator*. Reich and others used the orgone energy concentrated with this aid to therapeutic ends, irradiating patients with the energy and, as a result, obtained astonishing therapeutic successes. Over and above this, Reich carried out extensive research in the domains of medicine and biology, investigating the con-

sequences of this energy for the functioning of life processes and, in part, for the genesis of psychic, psychosomatic, and organic illnesses.

The Orgone Accumulator:
Principles and Mode of Operation

The principle of how the orgone accumulator works, considered by Reich to be the accumulation of *cosmic life energy*, can be discussed in simple terms. Reich discovered that insulators absorb this energy, while metals also pick it up, but then immediately reradiate it. According to Reich, using these properties will allow the energy to be accumulated in an enclosed space. If the walls around this space consist of a metal layer inside, and a layer of insulator outside, then the accumulation effect can be made stronger; additional layers make the accumulator stronger.

Reich used primarily accumulators designed to irradiate the entire body (for construction principles, see Fig. 1), so that the whole organism could be charged with life energy. A profusion of psychosomatic illnesses can be

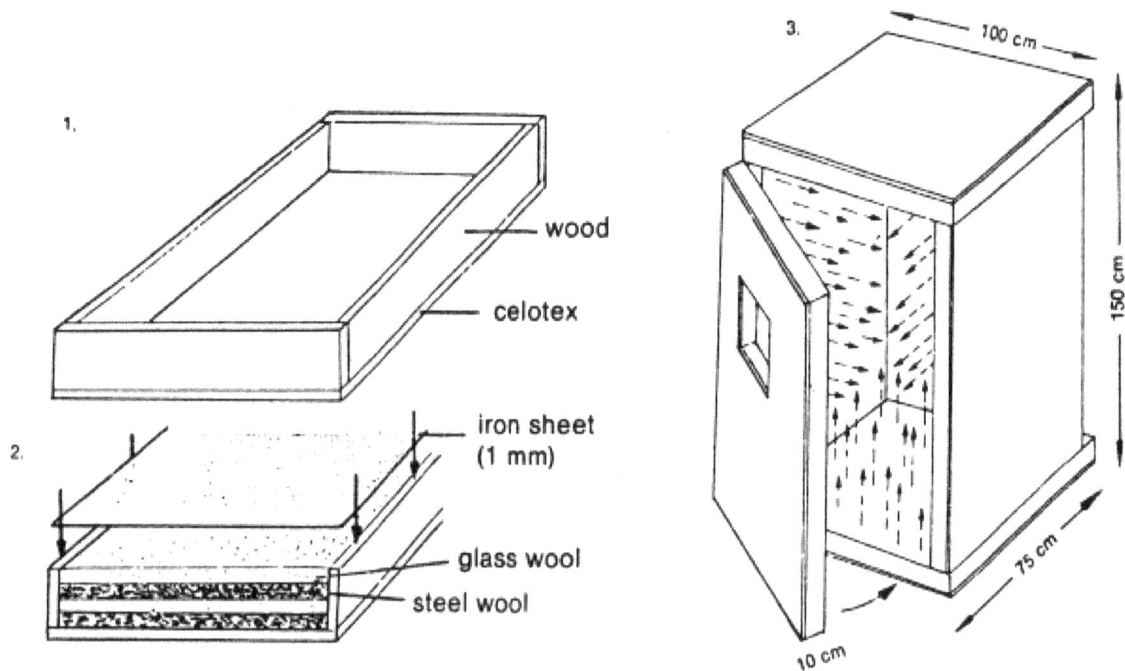

Figure 1.
Orgone Energy Accumulator for Body Irradiation.

cured—if one follows Reich's accounts—only through *regular* irradiation. In addition, the body's resistance to infection should, as a result, be increased. Further local irradiation of wounds and burns can contribute to a more rapid and less painful healing process. The accumulator, though, is on no account a panacea, mainly because its effects are too undifferentiated and non-localized. Here acupuncture seems to me to offer a more specific procedure. I will now come to some of the irradiation experiments that I have conducted and which, in my estimation, are suited to establish the identity of orgone energy as the energy which is the basis for acupuncture cures ("acupuncture energy", for short). In my view, these experiments open up new perspectives for a scientific foundation of acupuncture.

Special Orgone Accumulator for Irradiation of Acupuncture Points

If my hypothesis as to the identity of acupuncture energy and orgone energy is correct, then, I reasoned, it must be possible to expose single acupuncture points to concentrated orgone energy by means of a special, focused orgone accumulator, without needling or even so much as touching the points. In this way it should be possible to arrive at the same effects as are achieved with traditional needle acupuncture. I thus developed a special orgone accumulator that concentrates the energy on a small area—unlike the accumulator that Reich used—and thereby make it possible to irradiate just the acupuncture points.

The construction is incredibly simple (Fig. 2). Around an iron tube, 30 cm in length and 1 cm in diameter, I rolled about forty alternating layers of aluminum foil (metal)* and cellophane (insulator, also used as kitchen wrapping), with the outermost layer being of cellophane film. I reasoned that orgone energy, accumulated within the tube, would be radiated from its open end, and could then be directed at acupuncture points.

The hypothesis that acupuncture effects can be obtained by using this sort of radiation has now been confirmed by irradiating approximately one hundred people with it. Prior to application of the energy, a standard acupuncture pulse diagnosis is always carried out, and, moreover, only the tonification points of those meridians that exhibit an energy deficit according to the pulse diagnosis are irradiated. With symmetrical pairs of points, in each case, irradiation began with one point and then, after five to eight minutes, was shifted to the corresponding point on the other half of the body. In the course of five to fifteen minutes of irradiation, bodily sensations appeared that were quite definite and, in some cases, even disconcerting. It should be noted that persons treated had their eyes closed throughout, and were unaware as to the points the orgone tube was aimed at. They also had no knowledge of the position of the acupuncture meridians on the body.

---

* Reich warned against using aluminum for constructing orgone accumulators because the results of irradiation were dangerous to health. For medical purposes, applying iron (steel) has shown the best results. I, myself, have not had significant negative experiences by using aluminum foil, possibly because each treated person was irradiated only once; or possibly because the interior of the orgone accumulator tube was made of iron. Still, it is best to use a fine net of steel wire for constructing orgone tubes.

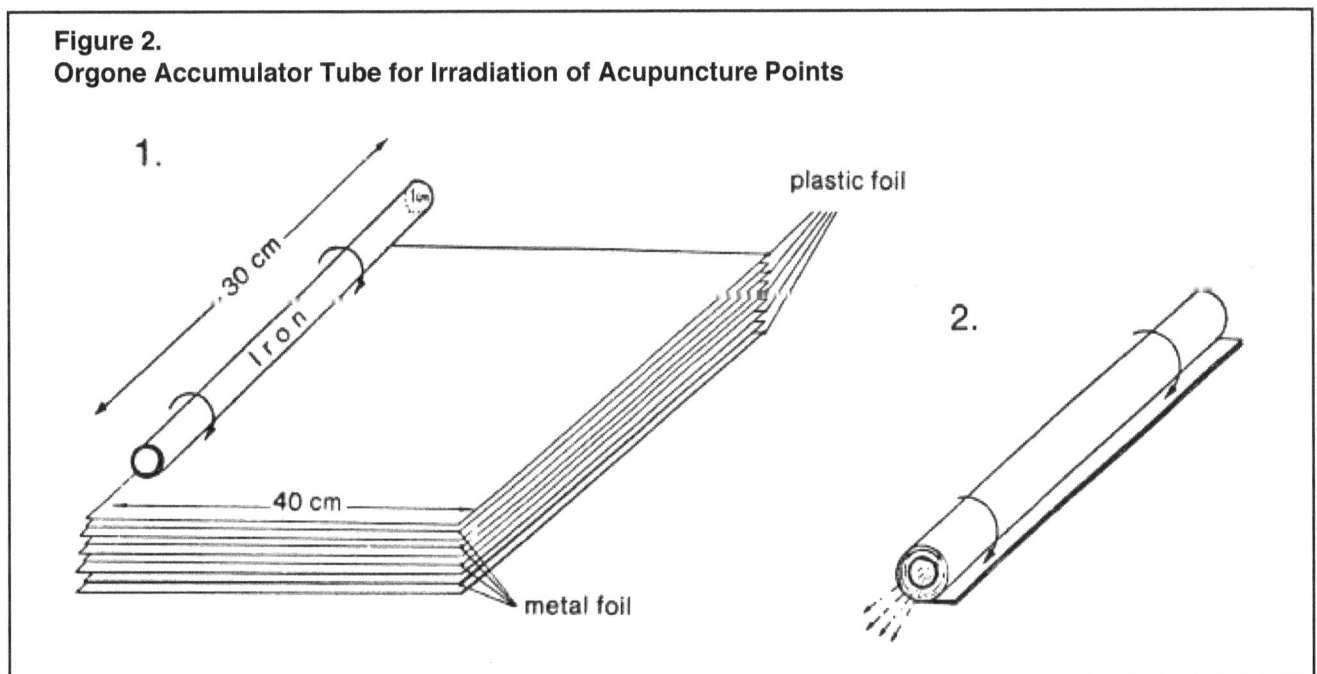

Figure 2.
Orgone Accumulator Tube for Irradiation of Acupuncture Points

For 85-90 percent of the people, the bodily sensations were undoubtedly reactions to the irradiations. For some people, the connection between irradiation and bodily sensation was not clearly established. A few perceived no noticeable effects either during or after irradiation. In about ten percent of the cases, overwhelming bodily sensations and some emotional breakthrough occurred.

Results of Acupuncture Point Irradiation With Orgone Energy

The irradiation of the little toe (acupuncture point BL-67, or tonification point of the bladder meridian) resulted primarily in a tingling sensation in the little toe, as if an electric field was being applied at the irradiated point, and in a warm streaming, felt first in the foot, then in the lower part of the leg, rising sometimes to the upper thigh, and even further, as a rule into the half of the torso on the side where the point had received radiation. Only a few people reacted specifically on the other side. Some subjects felt as if they were attached to a battery, or as if their leg had gone to sleep (though without the appearance of any other disagreeable sensations). In some cases they had the impression that the foot, or even the leg, had become gigantic, and there was often the feeling of a pleasant heaviness in the half of the body that had been irradiated. Generally, the streamings that appeared were described as extremely pleasurable. With several people there were spontaneous and uncontrollable spasm of the eye musculature (the bladder meridian originates precisely there, above the eyes). Sometimes strong streamings arose in the region of the bladder, and it felt like a blown-up balloon. Two persons, as a result of the irradiation, a short time thereafter experienced some emotional breakthrough in the form of sobbing (afterward this was felt as a release).

Irradiation of point TH-3 (tonification point of the Triple Heater meridian) on the back of the hand usually produced tingling sensations and warm streamings in the hands and then in the arms, often extending into the upper arm and into the area of the temple (the meridian ends in the temporal region). Sometimes feelings of giddiness appeared in conjunction with this irradiation, but more often there was a feeling of going round and round, as on a merry-go-round. In one case, the direction of rotation was sensed as changing shortly after radiation was transferred from the left hand to the right. With one person, who suffered from chronic pains around the heart, energy became blocked in the upper left arm; this was felt as disagreeable and painful. In another case, a woman experienced warm streaming and pulsing feelings in her vagina for the first time, and with these, strong pleasur-able sensations.

These same radiation-caused effects can be obtained by means of an orgone accumulator *plaster* that I have developed (see Fig. 3). The radiation strength of such a plaster, attached with adhesive tape over an acupuncture point, can be varied as much as desired by altering the number of layers of metal and insulating foil. With forty-fold accumulator *plasters*, effects very similar to those described here appear. It is my view that, with the possibility of a specifically localized dosage using various strengths of the orgone accumulator plaster, new therapeutic perspectives are opened up.

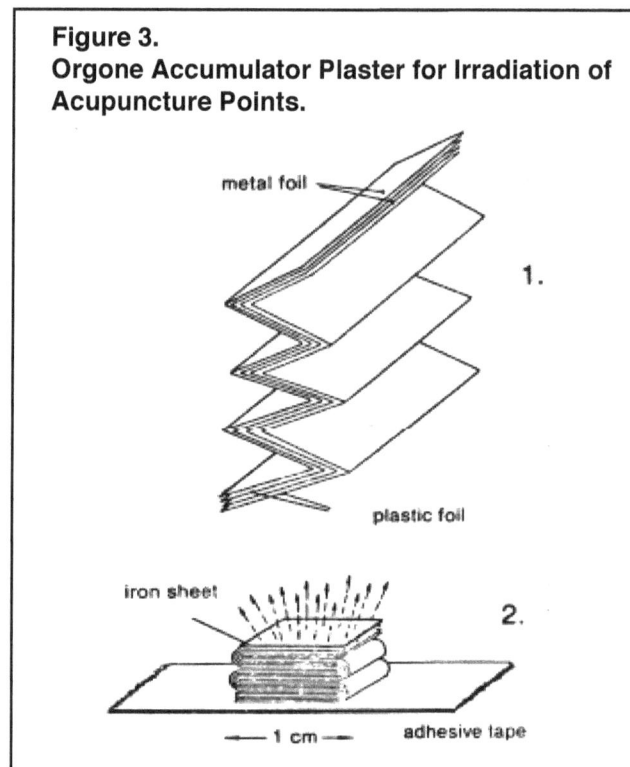

Figure 3.
Orgone Accumulator Plaster for Irradiation of Acupuncture Points.

Withdrawal of Orgone Energy from Acupuncture Points

It is also possible with the help of another device to draw orgone energy out of the body (again, without touching it). On the basis of what Reich called conformity to Natural Law, I have built an apparatus designed to draw energy from specific points. This apparatus consists of an iron tube, 40cm long, to whose end is soldered an insulated, braided-wire electrical cable, leading into some water. In the water, the wire should be stripped of its insulating sheath (Fig. 4). According to Reich, water strongly attracts orgone energy, so that in this way a sort of energy-suction is created. Because of this, the metal tube should never to be held with bare hands, since energy is drawn out of the hands of whoever handles the tube, in an uncontrollable and undesired way. As a result there can arise—as I have experienced—considerable feelings of lassitude and sharp pains in the body, as well as muscle spasm and strong emotional outbursts at night. Therefore, it is advisable to wrap a dry cloth several times around the tube, or better still, to fasten it to a stand and keep well away from it.

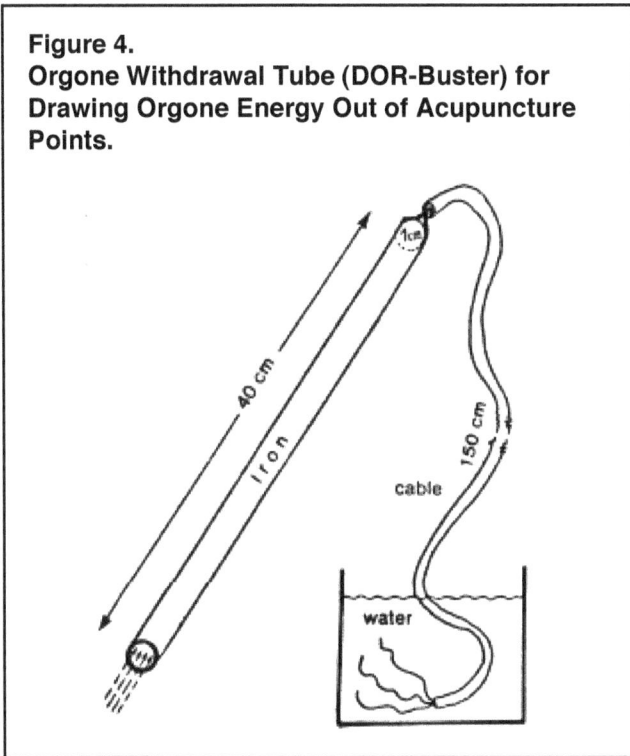

Figure 4.
Orgone Withdrawal Tube (DOR-Buster) for Drawing Orgone Energy Out of Acupuncture Points.

Application of this apparatus to the sedation points of overcharged meridians in the treatment of 50 subjects led to almost more amazing findings than did the orgone irradiation. The subjects felt in most cases a definite cool breeze at the sedation point treated, and had the feeling that "something was being sucked out" of them. In some cases, sharp pains appeared along definite pathways in the body, and when this happened, the paths coincided exactly with the course of the meridians whose sedation point was precisely that being treated. Here, as before, the people had no idea of the course of the meridian, nor in particular of the exact points being treated. Sometimes, cold shivers developed over the whole body, as if a cool wind were blowing around it. For others, localized muscle cramps were dissolved, and this was felt as agreeable. Particularly striking bodily sensations and reactions resulted in some cases where the heart and lung meridians were energetically overcharged and the subjects suffered from pressure and pain in the area of the chest. Here the

withdrawal of energy at point Lung 5 (sedation point of the lung meridian) sometimes led to strong spasms in the upper arm, shoulder, and chest areas, and to pleasurable feelings of release there. Breathing in the chest became easier and deeper, and remained that way, even after treatment.

Withdrawal of surplus energy from the liver meridian (point LIV-2) in some cases led to a spontaneous deepening of the breath, and even occasionally resulted in hectic abdominal breathing, with associate muscle spasms, particularly over the whole abdominal area. One treatment in which both orgone irradiation and energy withdrawal were employed turned out to be especially striking. First of all, overcharging the bladder meridian (BL-67) caused an emotional breakthrough in the form of sobbing (see above). At the same time oppressive pain was felt in the abdominal region. On then withdrawing energy from the liver meridian (point 2), the pain there was suddenly reduced, to be succeeded by enormously deep abdominal breathing and then chest breathing, with—through the effects of hyperventilation—tremblings throughout the whole body, but especially in the face, where they caused the teeth to chatter. Pulsing waves (blocked in the area of the pelvis) moved over the abdomen. While the irradiation, and in particular the withdrawal of energy, only lasted for about ten minutes, the body produced—gave itself over—to the reactions that I have described for about half an hour. After repeated outbursts of sobbing and imploring calls for boyfriend or mother, these emotions suddenly switched to completely free laughter and an overwhelming feeling of relief and comfort.

Conclusions

The experiments that I have conducted, and which are only sketchily presented here, allow no doubt—or so I believe—that acupuncture is based on purely energetic processes, and not on imagination or stimulation of the nerves, brought about by the mechanical procedure of inserting needles. In addition, it seems to me that these experiments demonstrate that the energy underlying acupuncture is identical with the orgone energy that Reich discovered and objectified. It is of utmost importance, in order to further these investigations, that the interest and cooperation of qualified acupuncturists be obtained, and some collaboration between acupuncturists and orgone therapists is established.

As additional topics for further investigation, the following may be considered:

- Objectification of the effects on the organism of point irradiation with orgone energy and also of orgone withdrawal (analogue of methods used in objectifying acupuncture effects) with appropriate control experiments.

- More detailed investigations of the therapeutic effects and possibilities of applying point irradiation with orgone energy and orgone withdrawal, by means of accumulator tube, accumulator plaster, and withdrawal tube, respectively.

- Continuation of Reich's physical and microbiological experiments, and a reexamination of his interpretation that these are explainable in terms of an energy form unknown until now.

The relevance of such investigations can, in my opinion, scarcely be overestimated. Through incorporating Reich's research findings, completely new perspectives for a scientific explanation of acupuncture may be opened up. Further, significant therapeutic possibilities may, as a result, perhaps be revealed, as the dosage of energy supplied or withdrawn by means of the orgone point applicator and "exhaust" can be so easily varied (according, respectively, to the number of layers wound around the tube, and to the length of the tube employed). Reich's findings could also be used to amplify the effects of acupuncture using needles, in that—in the case of tonification, anyway—not only could acupuncture needles be inserted in the patients body, but, in addition, he could be seated in an orgone accumulator designed to irradiate the whole body. By this means, the accumulated orgone energy could be directed and focused via the needles.

My experiments strengthen my supposition that the system of acupuncture is a refinement of the connection Reich discovered between pathology and energy flow in the organism. On the other hand, the teaching of acupuncture lacks any physical-scientific foundation, nor do they provide any biophysical-medical elucidation of the genesis of diseases. Both of these deficiencies could be overcome through an incorporation of Reich's extensive investigations. Moreover, Reich provides a profound analysis of those cultural mechanisms that—by creating environments and ways of bringing up children that are inimical to growth and emotional well-being—are the prime cause of energy blocks in the individual, as well as being the foundation for mass disease. The association of these two fields—acupuncture and orgone energy research—opens the way not only to extensive therapeutic possibilities, but also to far-reaching scientific and philosophical implications. Thus it is to be hoped that acupuncturists and orgone therapists, and particularly orgone researchers, will cooperate in working together in order to extend the experimental work that has now been initiated.

General References:

Boadella, D.: *Wilhelm Reich, the Evolution of His Work*, Vision Press, London, 1972.

Raknes, O: *Wilhelm Reich and Orgonomy*, St. Martin's Press, NY, 1970.

Reich, W.: *Character Analysis*, 3rd. Edition, Farrar, Straus & Giroux, NY, 1961.

Reich, W.: *The Discovery of the Orgone, Vol.1, The Function of the Orgasm*, Farrar, Straus & Giroux, NY, 1972.

Reich, W.: *The Discovery of the Orgone, Vol.2, The Cancer Biopathy*, Farrar, Straus & Giroux, NY, 1971.

Reich, W.: *Selected Writings, An Introduction to Orgonomy*, Farrar, Straus & Giroux, NY, 1960.

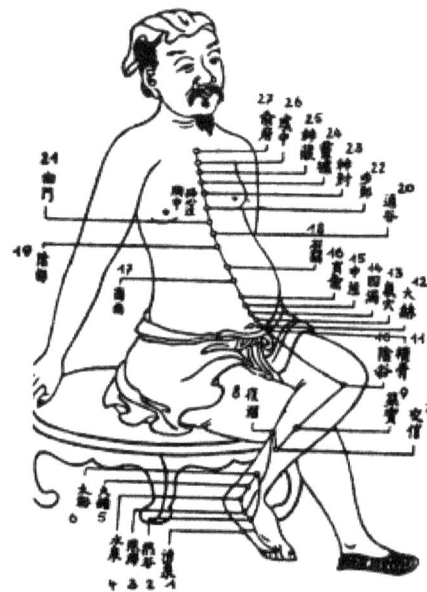

CONFERENCE REPORTS

Over the last several years, the Editor had the opportunity to speak at several major conferences and meetings, and talk with a wide variety of people on various aspects of orgonomic research. This section will informally present some observations made on the road, in the U.S.A., and abroad, for the readers of the Pulse. - J.D.

The Acres, USA Conference on Ecological Agriculture:

In June of 1986, the farm journal *Acres, USA* ran an article by Dr. Jesse Schwartz titled "Orgone Energy". The article was a straightforward report on methods for plant growth enhancement via use of concentrated orgone energy from the orgone energy accumulator. It had attracted my attention as it was, to my knowledge, the first *popular* magazine article to ever discuss Wilhelm Reich and the orgone energy in a factual, honest manner (other than those published by *Offshoots* or orgonomic research journals). I had been a subscriber to *Acres* for many years, given its straight reporting of environmental issues and other matters related to sustainable, ecological agriculture.

After reading that welcomed article, I wrote to the editor, Charles Walters, Jr., commending him for publishing it, and pointed out other aspects of Reich's works which were important for agriculture. Walters soon asked if I would come to Kansas City and give a presentation at the forthcoming *Acres USA* Convention. I agreed to come, and *Acres* later published a telephone interview with me on the subject of Wilhelm Reich's discovery of the orgone, and also a short article summarizing some of the atmospheric research I had done.

At the conference, I presented a paper on the "Agricultural Applications of Wilhelm Reich's Discoveries". It was truly an agricultural conference, dominated by farmers and their wives, with feed caps and coveralls. There were about 600 people in attendance, from all across the USA, but mostly from the Great Plains and Midwestern states. It was a refreshing change from the suit-and-tie environments I had experienced at academic conferences. These farmers looked you in the eye, and said what was on their mind. They were a select group, however, farmers with a concern for life and the land, for preservation of the soil, wildlife, and nature.

Besides focusing upon ecological agriculture, *Acres, USA* carries articles and advertisements on various "low energy", "vibratory" or "radionics" instruments for the purposes of increasing agricultural productivity, or for protection of crops from pest insects. Various dowsing instruments were also widely acknowledged and in use. Conceptually, these farmers required no introduction to the idea of a life energy, or an interconnecting energetic medium such as the orgone energy. They not only felt their kinship with the land, but had been working with various instruments to extend and enhance their perception and use of the life energy.

A large room at the conference hotel was devoted to displays by companies who sold such instruments. None of these individuals were selling orgone energy devices, or books by Reich, but in talking to the various presenters, I found most had been aware of Reich's work for a good number of years. Almost everyone I spoke with who knew about Reich acknowledged his work on the life energy as being very important. However, many were clearly uncomfortable with his work on adolescent sexuality and genitality, and their discussions about "energy" did not encompass emotion and genital pleasure. A few of the farmers, who had come from Bible-belt regions, were openly hostile to Reich's findings on adolescent sexuality, and there was a tendency to view the orgone in the terminology of radionics, or the ethers of Steinerian mysticism. However, these individuals were clearly a minority, and certainly can be found at all other forms of group meetings, not just among ecological farmers. About 50 copies of my Orgone Accumulator Handbook were sold within two days at the conference book table.

My talk proceeded on Reich and the life energy, giving a thumbnail sketch of his work on the emotions, and subsequent bioenergetic and physical research, to include a bit on the accumulator and atmospheric orgone energy. I showed a series of slides, to include about 10 with various manifestations of the blue orgone energy: the blue sky, blue mountains, blue oceans, blue aurora, blue Chrenkov radiation, blue cyclotron beam, blue bear wallow, blue Earth-in-space, blue energy field of a finger-tip, and blue energy field of an astronaut walking on the moon.

Questions were then taken from the audience, and these were directed towards specific information which they could use to help them in their work and daily life. A good number wanted instructions for building small accumulators to charge their seeds. Others wanted to convert their large metal grain storage silos into huge accumulators, to see if they could enhance the seeds during wintertime storage. Several others prefaced their questions with: "When I built my orgone accumulator...", and specifically asked about materials and construction techniques. In one case I could not meaningfully answer a question about accumulator construction, so another farmer raised his hand and said "I've tried that, and the best thing to do is..." and so on. I was pleasantly astounded. Here, I had come to bring new information to

these rural folks, assuming they would know little about it, only to find that Reich's books had made their way into their homes over the previous several decades. My only concern was that several of the farmers had built and experimented with primitive cloudbusters, and clearly had not taken the kinds of precautions necessary for safe or effective operations. Very few of the people I met knew anything about the various orgonomic publications or organizations.

Several of the conference participants later told me about their personal experiences with the orgone accumulator and blanket. Some had experienced marvelous benefits, eliminating colds and flu, and a variety of other minor complaints. A few others had built accumulators but did not experience any relief from their complaints, or actually experienced intensification of a particular symptom. In every such case I heard, the problems were related to an absence of specific information about proper materials for construction of the accumulator, or use of the apparatus in a contaminated environment, near to a television set, fluorescent lights, and so forth.*

One rather pale and low-energy fellow said he had cured his liver cancer with an accumulator. He said his doctor had diagnosed the cancer, and told him to "put his affairs into order" as he would last only a few months longer. Upon hearing this, he immediately began a program of detoxification, using a special diet and tea, and a several-times per day use of a home-built orgone accumulator. His accumulator was constructed from two 50 gallon steel oil drums welded together, sandblasted down to bare metal on the inside, and wrapped with layers of steel wool and fibreglass. He would crawl inside and pull the metal lid shut, leaving enough space for good ventillation. This fellow openly disagreed with my caution for people not to sit inside an accumulator for more than 45 minutes, saying that one time he fell asleep inside his accumulator for 7 *hours*! I told him that his low-energy condition appeared to have minimized the dangers, but that it still would not be a good idea for others to try. In any

---

* A new book by Dr. DeMeo, *The Orgone Accumulator Handbook* is now available from Natural Energy Works, Ashland, Oregon, USA. www.naturalenergworks.net

regard, his doctor became upset when he did not die, and especially when the tumor no longer showed up on x-ray films. The doctor accused him of taking some kind of "wonder drug" from another doctor, which he denied. When he told his doctor about his home treatment method, he was accused of lying, whereupon the farmer stopped going to the doctor altogether. All of this happened in 1984 and 1985, and the people in his small midwestern town were astonished to see that he did not die as predicted by the town's most reputable doctor. Accordingly, word about the "miracle box" began to spread throughout the town, and several of his neighbors have since started building their own orgone accumulators.

I had the opportunity to see this same gentleman in Summer of 1987. He was then more highly charged and mobilized than at our first meeting, had gained about 40 pounds, and had a ruddy, almost red-faced appearance. His increased energy level also expressed itself characterologically, as he had lost his low-tone, soft-spokenness, and now was highly animated, talkative, and often domineering. His internal tension had also built up rather quickly, and he was exhibiting signs of genital anxiety and the previously observed avoidance reaction regarding the accumulator. For example, he voiced a disagreement to me about Reich's work on genitality, and simultaneously seemed reluctant to attribute as much of his recovery to the accumulator as before. All of which underscores that the accumulator may greatly help and relieve certain symptoms an individual may have, but cannot by itself eliminate the emotional components of the more deeply rooted biopathy. In spite of this difficulty, use of the accumulator may have given him many additional productive years of life. Time will tell.

All in all, I came back from this conference with a greater appreciation for the growing interest in Reich's works. People very far removed from the big noise of city life are rediscovering orgonomy through the printed word, and in a small way putting it to practical use. In my view this is a very favorable turn of events, and not especially surprising. Unlike the educational ("teaching") or health care ("sickness") professions, which are highly organized into big institutions, farmers often retain autonomy and control over their work. The methods they use to grow a crop are not dictated by a central authority, and the police are not called out to jail the organic farmer (as is sometimes the case with a home-schooled child, or the healer who uses natural methods). There is no AMA, FDA, School Board, Principal, or Teacher's Union to answer to regarding farming methods. Farmers are also confronted with the sexual expressions of life energy in plants and animals on a daily basis; they must, if they are to survive economically, assist and guide such forces towards their natural ends in the development of fruits and progeny. Many farmers have chosen to use very anti-life, destructive methods to grow their crops or farm animals, but journals like *Acres* are increasing the awareness of farmers to the uneconomical and damaging character of such farming methods. Many farmers have already abandoned the damaging techniques, (eg., pesticides, herbicides, synthetic fertilizers, faulty plowing methods, etc.) in favor of those techniques which enhance the vitality of their soil and crops.

As a group, the farmers at the *Acres* conference convinced me of their strong positive feeling for nature and the life energy, and their independence of character. The ecological farmers may be one of the first large professional groups to apply, live, and openly acknowledge some of the principles of orgonomy. (Contact: *Acres, USA,* PO Box 9547, Kansas City, MO 64133)

The 11th International Congress of Biometeorology, Purdue University, Ft. Wayne, Indiana.

In 1987, at the invitation of Dr. E. Wedler of the Meteorological Institute of the Free University of Berlin, I presented a paper at the 11th International Congress of Biometeorology, which was held at Purdue University. That meeting was rather unusual as it constituted the first time in many years that the working members of the European "Piccardi Society" would get together to discuss research matters related to the findings of the late Giorgio Piccardi. For those readers who are unfamiliar, Piccardi was the Italian scientist who discovered an energetic principle at work in the natural world that permeated space and the cosmos, was responsible for solar-terrestrial interactions, which played a primary role in water "structuring" and weather phenomena, and was also reflectable by metal shielding. There is a similarity between some of Piccardi's findings and Reich's findings on the orgone energy. In any regard, Dr. Wedler had been a student of Piccardi, and we had exchanged a few letters on the relationship between Piccardi's and Reich's findings. After being invited to speak, I proposed to give a paper "On the Question of a Dynamic Biological-Atmospheric-Cosmic Energy Continuum: Some Old and New

Evidence", which would discuss these relationships, as well as some other relevant energetic discoveries. (Reprinted in this issue of the *Pulse.*)

The meetings at Purdue University were hosted by the International Society for Biometeorology, whose American branch is a fairly narrow group that does not stray very far from mechanistic concerns. Most of the American researchers were focused upon thermal or humidity stress upon one or another plant or animal. There were about 500 persons in attendance at this conference, mostly academics, and definitely a suit-and-tie crowd. About ten different meetings were held in different conference rooms simultaneously. One of these meetings was devoted to the Piccardi group, which was composed primarily of Europeans who were doing research on all kinds of fascinating subjects. Some were researching the electromagnetic aspects of clouds and storms as a means of weather forecasting, a subject known as "sferics". Sferic signals are radiofrequencies in the 15 - 500 kilocycle range, and violent storms are known to give off strong sferic signals. Normal clouds yield sferic signals of much lesser intensitites. This subject was once studied by American researchers, but was dismissed when a network of very crude detection instruments did not prove of sufficient value in the prediction of tornadoes. In Europe, with the coming of the computer and more sophisticated measuring instruments, sferics research has progressed to the point that maps of sferics emission across broad regions are being made, much in the manner of Radar maps which show regions of rainfall. Unlike the radar maps, however, the sferics detectors are totally passive, and do not require the sending out of a special "beam". This work was being undertaken primarily by Dr. H. Baumer, who has written a book in German titled "Sferics", and Dr. J. Eichmyer, who presented at the conference. Others at the conference were investigating the effects of low-level electromagnetism of different frequencies upon the living system. Others still were following up on Piccardi's work, in particular investigating the special "Piccardi box", which is similar in construction to the orgone energy accumulator.

As things turned out, I was the only American to present a paper in that conference room. I attributed this to the censorship of Reich's work in the USA, and the continued narrow-minded hostility which routinely punishes young American researchers when their work leads them in bioenergetic directions. I focused upon the relationship between Piccardi and Reich, noting the similarities between the construction of the orgone accumulator, with its layers of metals and non-metals, and the Piccardi enclosures, which consisted mainly of a metal box with an outer wool layer. In my talk, I pointed out how Reich discovered that the accumulator would yield anomalous results for various experiments run inside it, as

Final Announcement
1987

The International Society of Biometeorology

in conjunction with

American Meteorological Society

18th National Conference of
Agricultural and Forest Meteorology
and the
8th National Conference of
Biometeorology and Aerobiology

announces the

11th International Congress of Biometeorology

September 13-18, 1987

at

Purdue University
West Lafayette, Indiana 47907
USA

compared to outside of it. I discussed the To-T experiment, the electroscopical discharge rate differential and water evaporation rate differential experiments... and observed that nobody in the audience was even blinking or uncomfortable, which is usually the case when Reich's findings are presented to an unfamiliar group of scientists.

Actually, the Piccardi researchers had been observing similar unusual results from their metal box experiments, mainly the influence upon the physical chemistry of water. In particular, Piccardi and his co-workers observed that distilled "super-cooled" water would express a patterned alteration in freezing temperature, which would vary from day to day, depending upon cosmic and meteorological factors. Chemicals dissolved into water would likewise develop patterned changes in their precipitation rates, showing that a good portion of "standard experimental error" was actually regulated by a cosmic, energetic principle which was capable of penetrating metal enclosures and glass test tubes. Indeed, the Piccardi researchers had some 40 years of published evidence, primarily in the Italian and German languages, corroborating these very kinds of experimental results: different results obtained within the metal box than outside of it; different results in rainy versus clear weather; different at one season of the year as opposed to another,

but in all cases differences of a very lawful, patterned nature.

Furthermore, these researchers had corroborated one of Reich's earlier findings: the results of experiments performed inside the metal Piccardi box would be disturbed by nearby fluorescent lights or televisions, and proceeded best when performed in a very natural environment, preferably one of trees and natural surroundings. In short, the work of these Europeans had corroborated many aspects of Reich's findings, but in a penetrating manner that in some respects had gone beyond Reich's own findings, particularly regarding the physical chemistry of water. In fact, the only time my European hosts seemed amazed at my own presented work, was with the discussion of the cloudbuster, where they learned that the phenomenon at work inside the accumulator (the orgone energy), could influence cloud dynamics at a great distance, through use of a hollow metal tube grounded into water. This latter observation was, of course, clear proof that the phenomenon was not one of an electromagnetic nature, but something quite different, of a more primary nature.

The Bear Tribe Medicine Wheel Gathering, Payson, Arizona:

In 1987, I began a correspondence with several members of a group called the *Bear Tribe*, which is based in Spokane, Washington. The group publishes an excellent quarterly magazine, called *Wildfire,* which focuses upon environmental issues, earth changes, and native American issues. Several articles appeared which discussed childrearing in a sex-positive and life-affirmative manner, articles speaking out about the hazards of circumcision, and about adolescent sexual freedom. A few articles also appeared by Jerome Eden, who presented Wilhelm Reich's work in a straightforward manner. Eden had a considerable amout of contact and influence with a few Bear Tribe members, and they always spoke highly of him. The magazine's editor, Matthew Ryan, once invited me to Spokane to give a workshop, and he assisted with a cloudbusting operation in that area, during a severe drought that gripped the entire Pacific Northwest (see the "CORE Reports" section of the *Pulse* for discussion on that operation). *Wildfire* later reprinted an article of mine, and I was invited to participate as a teacher at one of the Tribe's Medicine Wheel Gatherings. I accepted this opportunity, and in late September of 1988, drove from our headquarters in California to a wooded location deep in the Arizona mountains near Flagstaff. The other speakers at the gathering included Sun Bear, Wallace Black Elk, Grey Antelope, Brooke Medicine Eagle, and

Sun Bear, leader of the Bear Tribe

Dawn Featherhorse, who were all native American peoples. Additionally, two Hawiians were there, Kahiliopua and Auntie Lani. My name on the list of speakers, "Dr. James DeMeo", stuck out like a sore thumb. I also admit to some anxiety about speaking at a function which I feared might be steeped in mysticism. However, I resolved to do it, based upon my previous good experience with *Wildfire* and the workers at the Bear Tribe headquarters in Spokane. (To date, I still have not read any of Sun Bear's books.)

It first should be pointed out that most of the members and participants in the Bear Tribe are not native American peoples themselves. Only the leaders of the Tribe are, such as Sun Bear, who sports shoulder-length black hair, and in prior years has acted in a few films. Many of the young people working for the Tribe are from the large cities of America, having left bad situations at home, or are simply in retreat from the destructive tendencies of our "civilized" culture. Some of the people working with the Bear Tribe were undergoing the emotion-release orgone therapy that was developed by Wilhelm Reich, and were familiar with orgonomy. I was told that Sun Bear at first did not know what to make of this, but he eventually embraced the therapy as a good thing, because he noted that "it makes people behave more rationally". Sun Bear's

gentle philosophy of healing the planet, presented through Native American rituals at various Medicine Wheel Gatherings across the USA and overseas, has spurred a resurgence of interest in native American traditions. His is an animistic world view, which he believes is needed to bring people back to nature, back into contact with the life energy, in order to save humanity. And I agree with this position.

Once in Arizona, I was greeted most warmly, and met many of the native American teachers who regularly speak at the Medicine Wheel gatherings. They had much to teach about communion with nature, and their long and familiar tradition with what they called the "sacred energy". For example, I learned that the original native American words for "God" or "Great Spirit" had been mistranslated. The correct translation for some 300 different dialects of native American language indicated that the "Great Spirit" was not a noun, but a *verb*, with the meaning of "energy", or the "Force-That-Moves-All-Things". For example, the Ojibwa called this force *Gitchi Manitou;* to the Sioux, it was *Wakan Tanka,* and indicated the "Great Mystery which Penetrates and Forms the Universe". To the Otoe people, it was called *Waconda.* The Huichol called the energy *Kapuri,* and actually made yarn paintings of it, which showed a double helix pattern. These descriptions are in complete agreement with Wilhelm Reich's discovery of the orgone energy, or life energy, which was identified in a natural scientific context; the visible orgone units, which move in a spiraling manner, was noted by the native shamans.

Additionally, the term "Indian" was not from the error of Columbus, but from early Spanish priests, who saw the devoted respect for life and nature that permeated the daily life of the American natives, whom they called "In Dios", or "with God". As the Bear Tribe gathering proceeded, I observed that many customs of the native America peoples were based upon a very direct observation of life energy functions in nature.

A seriousness permeated many of the activities of the gathering, to include the creation of the medicine wheel, the sacred pipe ceremony, and the sweat lodge. At the sacred pipe ceremony, short prayers or wishes were offered up as sage-herb pipe smoke blown into the air by the participants. The native prayer itself, or the appeal to the cosmos, was literally "put into" the atmosphere, and spread into infinity, by the act of blowing out the smoke. The atmosphere, and not some heaven or nirvana in another twilight-zone dimension, was the residence and home of the Force-That-Moves-All-Things, and appeals were directed towards it through the agency of the smoke. For the record, no pot was in the pipe, and Sun Bear forbids his apprentices from using drugs or alcohol, which are likewise forbidden at Medicine Wheel Gatherings (one of the few rules). Indeed, there was no marijuana or peyote to be seen at the entire gathering, which had attracted a variety of young and old people, from all different ethnic and social backgrounds.

During the nighttime sweat lodge ceremony, large stones the size of coconuts were heated up to a glowing incandescence, in a roaring bonfire. At the appropriate moment, the stones were removed, and transported a short distance into a shallow hole in the ground, which was in the middle of the sweat lodge. About 20 of us were crammed into this round sweat lodge, which was only 10 feet in diameter and about 3 feet high. It was made airtight by many layers of blankets which lay across its open pole framework. When the door flap was shut, the stones began to bake you, and the sweat streamed. Songs were chanted and drums beaten loudly inside the sweat lodge, and then water was splashed upon the red-hot stones, steaming everyone until quite well-done. However, more than steam was at work in the sweat lodge. Reich found that red-hot sand, when immersed into a solution, would break down into microscopically glowing blue *bions*, liberating a great deal of radiating orgone energy. This principle appeared to be at work in the sweat lodge.

The man leading the sweat ceremony, Wallace Black Elk, affirmed this in his own way, informing us that the sweat lodge was a place of both physical and emotional healing. He told us stories of the physical healing of debilitating diseases in the sweat lodge, and also of people with "wounded hearts" who came into the lodge and either cried or raged, emerging with a whole feeling. The sweat lodge resembled an energy accumulator/ therapy session in combination! "The stone people come out of the rocks. They remind us of our ancestors and our youth, and you cannot hide your problems from them. During the chanting, you tell the stone people about your anquish and troubles, and call upon them for help. They will listen."

When water was thrown upon the red hot stones, and the chanting and drum beating resumed, many of those inside were driven to tears or anger, some weeping or shouting quite openly. This was both from the intense physical sensations involved, but also because of the feelings that were stimulated. I wish to put aside any comparisons here with "group therapy" sessions, and emphasize the rather private nature of the space inside the sweat lodge, where you cannot even *see* the person next to you, and where the mind is filled with flashing images stimulated by the intense heat, pounding drums, and chanted songs. No intellectual appeals were made for people to "let go"; the steam and hot rocks simply charged you to the bursting point, and the darkness and loud chanting provided the anonymity necessary for letting go of pent-up feelings. Was this one of the ways that native peoples regularly purged themselves of anxiety and armoring? Certainly, Black Elk and others at the Medicine Wheel *knew from experience* that native customs like the sweat lodge could help to reduce much of the chaos and destructive behavior on the reservations. Unfortunately, these social customs had often been banned by the laws of the white man. In my view, the sweat ceremony would be good for all people. As a means for getting people to let go of pent-up feeling, it is a far cry better than drugs or mental institutions. It makes the European sauna appear insipid and timid by comparison.

Another interesting ritual I learned about was performed by the apprentices to Sun Bear, who carries a message of "love Mother Earth". He advises his students to go out into the forest, find a place where they are all alone, and dig a hole in the ground about a foot deep and a foot across. Then, "you lie down on your stomach facing directly into the hole, and speak to your Mother Earth. You tell your Mother what is bothering you, tell her those things that you have been holding inside, and never told anyone else". People who have done this often report a great opening up, and welling up of feeling with deep crying, sobbing or rage, and a great emotional relief afterwards.

The "spirituality" of the native American, as exemplified by the above examples, was *emotional,* with ritualized methods of bringing buried emotion to the surface, and *releasing,* rather than damming-up internal tension. At almost every point, the native American belief system and its various rituals contrasts with the conceptual frameworks and rituals of what the anthropologists call the "High God" cultures: Christianity, Judiasm, Islam, and Hinduism. As I previously demonstrated in my research on the "Saharasian Connection", outside of a few very well known groups, the majority of native North American peoples were *sex-positive, child-positive, and peaceful.* What I experienced at the Bear Tribe Medicine Wheel gathering was a very pleasant and encouraging confirmation. (*Wildfire* magazine, and the Bear Tribe, PO Box 9167, Spokane, WA 99209)

The Congress on Geo-Cosmic Relations, International Committee for Research and Study of Environmental Factors, 1989, Amsterdam.

Following the very fruitful conference at Purdue University held by the International Society for Biometeorology, described previously, I received a new invitation to present my research findings at the First International Congress on Geo-Cosmic Relations, in Amsterdam, Holland. The Congress was hosted by the Foundation for Study and Research of Environmental Factors, and organized by Dr. G.J.M. Tomassen, of the Department of Ecological Agriculture at the University of Wageningen. The list of speakers at the Congress included a host of people whose research had influenced me during my student years, to include many European "biometeorologists" who had studied and documented cosmic energy principles at work in the natural world.

About 100 different scholars gathered for the Congress, and a short review of the list of participants revealed an amazing situation. While only 7% of the participants were from the United States, 69% were Europeans, while a full 24% were from either the Soviet Union or other East Bloc nations. It was clear that the energetic concerns were taken very seriously in Europe and the East, but not so seriously in the USA. My presentation to this group was essentially the same as was previously given at Purdue Univeristy: "On the Question of a Dynamic Biological-Atmospheric-Cosmic Energy Continuum: Some Old and New Evidence" (See p.1-9)

The presentations which were most impressive from a bioenergetic viewpoint were those by Dr. Eichmeier from Germany (on sferics), Dr. C. Capel-Boute of Belgium (on the Piccardi effect), and Dr. M. Gauquelin of France (on the cosmic effects upon human behavior and heredity). Gauquelin was one of the most interesting of the presenters, summarizing his extensive research on planetary effects upon human behavior, about which he has published books in both French and English. The Congress was, he remarked, the very first time he had been invited to present his work before an "academic" body, since his first publications in 1955! His research on subtle planetary influences, which corroborates a few of the precepts of astrology, has been the subject of vicious attacks over the years. One of the leaders of this attack, in the USA at least, has been the American CSICOP organization.*

Another interesting speaker was Dr. N.V. Udaltsova, of the Institute for Biophysics in Moscow (one member of a contingent from that Institute), who presented information on structured, non-random patterns in radioactive decay processes, suggesting that the atomic nucleus was under the influence of a very penetrating and powerful

---

* See the *Emotional Plague Notes* for more details.

outside force. This was the first time I had seen anything on the subject outside of Reich's work, and a very few American publications from the 1970s which appeared in physics journals. When she was asked about what she thought was at work to cause the interesting patterns, she replied that it was merely a mathematical aspect of the data, without any physical mechanism, in keeping with the theories on "chaos".

For the reader unfamiliar with chaos theory, it is a wide field which allows mathematical and computer models to closely duplicate the visible structure and behavior of natural, self-organizing systems, such as life functions and weather patterns. Chaos theory affirms the need for an interconnecting, organizing principle in nature, but it has been used by some mechanistic scientists to suppress discussion of any *tangible* or *physically measurable* organizing principle. Indeed, at least one of the American participants was keen to explain ("explain away"?) most of the unusual natural processes identified by the Congress researchers as being the result of "chaos" patterns in their data, which he felt required no tangible causal mechanism. This individual refused consideration or discussion of empirical evidence for the measurable and observable orgone energy continuum ("that's impossible!"), but he later gave a talk on the "chaotic ordering" of the universe by virtue of a mystical and undemonstrated "consciousness" factor, which expressed itself only at the "quantum level". My argument, that the causal force behind the Piccardi effect, the sferics phenomenon, and the Gauquelin effect might be aspects of the tangible and highly structured ("chaotic") cosmic orgone energy, was not well received by this small group of chaos enthusiasts.

This Congress exemplified the best traditions of science, in that there was rich and vigorous interaction among a group of scholars from around the world, who could contribute to each other's work in a constructive manner, and voice opinions and disagreements without fear of personal repercussions or censorship. (Contact: Dr. E. Wedler, CIFA: International Committee for Research and Study of Environmental Factors, Meteorological Institute, ~~Freie Univ. of Berlin, Dietrich-Schafter-Weg 6-10, D-1000 Berlin 41~~, West Germany). Note: CIFA will host another Conference in Vienna, Austria in1990; see the *Pulse* Calendar for details.

NOTE: The discussion on "Orgonomy in West Germany" that was originally scheduled for this issue of the *Pulse* will instead appear in the Spring 1990 issue, which will also carry Conference Reports from Japan, and from the First International Symposia on Circumcision, the annual Cancer Control Society Convention, and the annual Convention of the American Society of Dowsers.

Sugar Glider on Eucalyptus Flowers
by Beth Cook

NUCLEAR HAZARDS

❖ The Pacific island atoll of Bikini was used as a USA atmospheric nuclear bomb test site in the 1950s. Fallout from these tests dusted the nearby islands of Ailinginae, Rongelap, and Rongrik. Some 15 years ago, the Department of Energy studied the effects of this fallout, and found that the bones of the inhabitants of those islands were contaminated with plutonium, and other radioactive isotopes that had been taken up in the food chain. However, the findings were not made public until most recently. The people of Rongelap were showered with radioactive fallout dust following a Bikini hydrogen bomb test, on 1 March 1954. For 2 days they were exposed to extremely high levels of radiation, until finally evacuated. They were returned to Rongelap in 1957, when the US Government declared the island "safe", and lived there until 1985, when the Greenpeace ship *Rainbow Warrior* helped them to relocate to safer islands. This same ship was later sunk by French terrorists, after it successfully disrupted French nuclear tests. Now, the US Government wants the islanders to return to Rongelap, but they are correctly suspicious. A 1988 study found that the island was safe only "for adults", but the northern half was still considered unsafe for anyone to "eat or sleep" there. Scientists working at two US Department of Energy Laboratories (Brookhaven, New York, and Lawrence Livermore, California) are responsible for the 15 year cover-up of the radiation findings. (*Science*, 14 July 1989)

❖ A recent story in the *Greenpeace* magazine (July/August 1989) gave the history of various accidents at sea involving ships or planes carrying nuclear weapons, or powered by nuclear reactors. In many cases, nuclear bombs or nuclear reactors have been lost, down to the deep ocean floor, in some cases never to be recovered. Aircraft carrying nuclear fission materials have crashed at sea; nuclear tipped test rockets have malfunctioned and fallen into the sea; entire nuclear-powered submarines, often carrying dozens of nuclear warheads, have sunk; extremely radioactive coolant has been dumped overboard; and entire nuclear reactors have been dumped into the sea like so much garbage. Throughout all this, and in spite of dozens of reactor failures and accidents at civilian nuclear power plants, the Navy and the nuclear industry continues to repeat the following lie: "Over 3,500 reactor years of operation without a reactor accident".

❖ The Rocky Flats nuclear waste dump, near Denver, Colorado, has for years been accused by environmental and anti-nuclear groups of contaminating the local air, soil, and groundwater with radioactive garbage. Most recently, the federal government finally acknowledged that this contamination exists, and the FBI is looking into charges of illegal dumping, and falsification of records. Some of the contaminants recently confirmed by the Environmental Protection Agency include the deadly fission products of cesium and strontium.

❖ Radioactive debris from the Chernobyl nuclear disaster has recently been found in supplies of powdered milk in Bangladesh. The powdered milk came from dairy farms in Poland, having been put into the Asian marketplace without prior checking for radioactivity. Poland initially took back shipments of contaminated milk powder, but then suddenly ceased doing so, claiming that a shipment of tea from Bangladesh was also radioactive. (*Earth Island Journal*. Spring 1989)

❖ Fallout from the Chernobyl nuclear accident site is spreading north from the Ukraine into Byelorussia, even though it is three years after reactor number 4 exploded (in April of 1986). More than one-fifth of Byelorussia's cropland is now contaminated, with 408 populated areas requiring constant monitoring. (*San Francisco Chronicle*, 12 Feb. 89)

❖ Nuclear waste created by US nuclear power plants and nuclear weapons production still has no where to go, and is piling up fast. The Governors of Idaho and Colorado have both indicated that they will soon cease allowing new shipments of radioactive garbage into their states, and the various schemes for "permanent" disposal always seem to have environmental problems that scuttle them.

❖ Opposition is growing to the planned launching of space satellites that contain nuclear power plants on board. These "orbiting nukes" are much more compact and smaller than the Earth-bound variety, but pose a potential hazard that is just as great. Most of the small nuclear power "packs" contain dozens of pounds of radioactive *plutonium*. Should one of the satellites fall back to Earth and burn up, or should one of the launch vehicles explode on the way into orbit, this plutonium might also burn up and turn into a cloud of radioactive smoke, mixing widely within the atmosphere. Eventually, the tiny radioactive smoke particles would fall to the Earth and be breathed by humans, or be taken up into the food chain. Dozens of these orbiting nukes are planned to power the various "star wars" devices, as well as for various deep-space probes, such as the forthcoming *Galileo* mission to Jupiter. A citizen's protest rally, for "No Chernobyls in Space", is planned for the October 21 launch of the Galileo probe, which will carry *50 pounds of plutonium*. Previous protests have been held at the Fairfield, Connecticut headquarters of General Electric, makers of the plutonium power pack. Solar power, which could usurp the role of power companies here on Earth, appears to be "out of the question" for such space purposes. Opponents of nuclear power in space say that several tablespoons of plutonium is all that it would take to produce a guaranteed lung cancer for every person on Earth. Proponents of the nukes say that this is not true, that it would take at least *two pounds* of the stuff to kill everybody. *Not to Worry!* (Contact: Women's International Coalition to Stop Making Radioactive Waste, 77 Homewood Ave., Allendale, NJ 07401.)

❖ The problem with orbiting nuclear power plants was recently highlighted by scientists engaged in "gamma ray astronomy". These scientists use sensitive probes to measure solar flares and cosmic gamma-ray bursts. In recent years, however, their observations have been disrupted by the increasing use of nuclear power plants on space satellites. Each time a nuke passes overhead, the gamma ray telescopes are temporarily "blinded" by the relatively large amounts of radiation being emitted by the satellites. Count rates in the astronomer's gamma ray detectors can climb from two to five times the normal levels within a matter of minutes. (*Science,* 28 April 1989)

❖ Problems associated with radioactive materials in space date back to the 1950s, when "Project Argus" and "Project Starfish" were undertaken. These were typical "mad scientist" experiments which involved rocketing nuclear bombs high up into the Earth's atmosphere, and then igniting them, for no obvious reason other than to "see what would happen". What happened was that geomagnetic and ionospheric disturbances were registered world-wide, forming a new, highly toxic radiation belt around the Earth. According to one report "The starfish blast, in fact, created a... [radiation] belt that extended from low altitude to a distance of 19,110 km, three times the thickness of the Earth. The degree to which these tests may have disrupted the naturally-occurring radiation belts will never be known since, as the *Encyclopedia Britannica* notes: 'the electron fluxes have changed markedly since 1962 as a result of high-altitude nuclear explosions by the US and USSR'." A few astronomers protested the plans for these high altitude explosions, but they were ignored. The possible relationship between these irresponsible nuclear explosions and the deterioration of the ozone layer is likewise ignored. (*Earth Island J.,* Winter 1988-89)

❖ Genetic mutations were recently discovered in small animals living near the Fernald Uranium Processing facility near Cincinnati, Ohio. Reproductive problems and other mutations were found in animals living near the plant's incinerator and waste pits, which contain low-level radioactive uranium. (*San Luis Obispo Telegram Tribune,* 2/3/89)

❖ **Food Irradiation Alert**: The Food and Drug Administration has already approved the use of atomic radiation for so-called "food preservation" purposes. Worse, the FDA does not want you to know which foods have been treated, and which have not; they are resisting all attempts to force food manufacturers to label food products where irradiation is being used. This is a wildly irresponsible biological experiment being launched upon a sleeping public. *Watch out for the following symbol,* which means that the food upon which it is posted has been exposed to intense gamma radiation. (Contact: *Stop Food Irradiation*, PO Box 59-0488, San Francisco, CA 94159.)

❖ *Power Surge* is the first comprehensive study on the contemporary and potential uses of renewable energy since the Carter years. The report states that "the electric capacity of renewables other than hydropower has increased ten-fold over the past ten years; they now supply the equivalent of about eleven 850-MW coal or nuclear plants". The report indicates that renewable energy — hydropower, geothermal, direct solar, wind and biomass — today provides fully 8% of US energy needs, a figure that will grow to 19% by the year 2000, and possibly to 80% by 2010. The report documents the slashing of renewable energy funding (in favor of nuclear energy) during the Reagan years. The USA spends only 8% of its energy research budget on renewables, compared to 12% in West Germany, 22% in Sweden, and 63% in Greece. In 1980, the USA held 75% of the world market in photovoltaics; today, that figure is only 32%. Japan's share of the photovoltaic market jumped from 15% to 37% over the same period. (Available for $20 from *Public Citizen*, 215 Penn. Ave., SE, Washington, DC 20003)

❖ The 30,000 children living near the Hanford Nuclear Reservation (Dump) in Washington state during the 1940s and 1950s may have been exposed to more radioactive iodine than Soviet residents near the Chernobyl reactor accident site. So says a recent study by the Centers for Disease Control. (*Radioactive Rag*, April 1989)

❖ Over 3000 acres of the Nevada Nuclear Test Site are contaminated with radioactive or other hazardous wastes, according to a report released by the Department of Energy. Over 680 nuclear tests have been held at the site in the last 40 years, many of them above ground until 1963. There are 777 different sites with radioactive releases, according to the report. Government officials in Nevada are very upset about this report, as the same Department of Energy wants to have a blank check for developing a "safe" nuclear waste repository in the Yucca Mountain region. (*San Francisco Chronicle*, 5 August 1989)

❖ Little is being reported in the US or USSR of the growing antinuclear movement. Hundreds of protesters have repeatedly made public attempts to "reclaim the land" at our Nevada Test Site prior to announced nuclear detonations, but by reading only the newspapers, you would never know this. In the USSR, a growing movement of dissidents have likewise organized and made public calls for a cessation of testing. In both cases, the long history of environmental abuses of the testing programs have provided sufficient reasons: radioactive leaks, increased mortality and morbidity of nearby residents, destruction of fragile desert lands, and concerns about global effects. In the USSR, 5000 people showed up for one such protest in Moscow, and in sympathy with the activities in the USA, their group is called the "Nevada Movement". (*Earth Island Journal*, Summer 1989)

ELECTROMAGNETISM EXPOSED

❖ A recent series of articles focused upon the hazards of low level electromagnetic fields appeared in the *New Yorker* magazine (June 12th, 19th, and 26th, 1989) by columnist Paul Brodeur. The articles reveal the hazardous nature of the electromagnetic energy fields that are given off by electrical power lines, television sets, microwave ovens, fluorescent lights, computer video display terminals, electric blankets, and so forth. Even the manner in which local power lines and grounding wires are interfaced within a given home or neighborhood can cause biologically significant disturbances, which may be at the root of a number of wide-spread health problems. Childhood leukemia, for example, has been correlated to proximity to power line fields, while miscarriages occur more frequently to women who use electric blankets, or who work with computer video display terminals (VDTs). The articles also document the short-sighted, narrow-minded, and indeed, criminal nature of many government and industry officials, who generally have sought to downplay the hazards involved, or to blatantly cover them up. Brodeur will soon published the articles in book form (*Currents of Death: Power Lines, Computer Terminals, and the Attempt to Cover Up Their Threat to Your Health*, Simon & Schuster) Of interest is the fact that conventional theories of electromagnetism are not sufficient to explain the observed effects upon people and organisms. The effects from power line fields may involve the orgone energy field of the Earth, the atmosphere, and also of the affected organism as well. The magnetic component of the EM field, for example, behaves almost as if it were "alive" in the sense of the orgone energy.; the magnetic component of the EM field merges with the geomagnetic field, can go around barriers or completely through shielding, and may "disappear", only to reappear at a new location. As discussed in a paper (by DeMeo) in this issue of the *Pulse*, the EM field also requires an orgone-like "medium" through which it can transmit its influences.

FOREST ANNIHILATION

❖ While Brazil and other rainforest nations are being asked to cease cutting down their rainforests, to spare the global ecology, lumber companies and bankers in the USA are pushing hard to clearcut the last remnants of old growth rainforest in the Pacific Northwest. The small, northern spotted owl, which lives in these woods, was placed on the endangered species list most recently, sparing its woodland habitat from the axe. The forest service proposed that 374,000 acres of forest be taken out of production, which is

only 5% of the total timber supply in that region. However, there remain only 6 million acres of old growth forest in the Pacific Northwest, down from 19 million acres present when the first white settlers arrived. While lumber companies push to clear cut as much land as they can, environmentalists and lumbermen have clashed in rival protests at clearcutting sites. Anybody who has traveled on the backroads of northern California, Oregon, or Washington will not have missed seeing the areas devastated by clearcutting. Contrary to popular myth, little replanting has been done on federal lands, where timber is sold off to lumber companies at bargain-basement prices. A large percentage of this wood, and the wood taken from Alaska, is shipped to Japan, where it can be sold for a better price. Lumber mills in the Pacific Northwest have been hit particularly hard, as the whole logs are generally shipped overseas, without first being milled into lumber. Decades of abuse by lumber companies, and lack of concern by other citizens (who care only for wood at a cheap price), has led to the current crisis situation.

❖ Japan has entered into contracts with the government of New Zealand, and is now in the process of clearcutting one of the last remaining stands of old growth forest on the North Island. In a related move, the hardwood rainforests of Malaysia (Sarawak-Borneo) are being turned into chopsticks for Japan, and cheap plywood for the USA. Logging here is happening at twice the rate as in Brazil, and will only last another 10 or 20 years at most. These nations will reap a short term economic boom from the clearcutting, after which all the environmental disasters associated with deforestation will occur: massive soil erosion, stream siltation, extinction of forest species and fish, atmospheric stagnation and drought. This says nothing about the deadly effects upon the native peoples who have lived in these forests for centuries. In Malaysia at least, native land rights are not recognized, which has led to protests, and blockading of logging roads. Government troops were called in on one such occasion, arresting 43 Penan leaders and clearing the way for loggers. A new Forest Ordinance decrees a two-year jail sentence and a $2,500 fine for anyone obstructing logging activities, and the costs

for dismantling of blockades can be charged against funds held by native tribal groups. Environmentalists in Malaysia have also been arrested and subject to stiff fines and penalties, evidencing a clear collusion between businessmen, loggers, banks, and government. "Thailand was once the world's top supplier of hardwood. Then their forests disappeared. It's now a net importer of wood. The Thais learned their lesson the bitter way. First they banned timber exports. Now they've banned logging altogether." (*Earth Island Journal*, Spring 1989; *Rainforest Action Network*, September 1988.)

❖ Spruce trees in the high elevation Bavarian forests, called the Fichtelgebirge, about 10 years ago began to suffer from an unhealthy springtime yellowing of needles. In 1982, the yellowing affected 8 percent of the trees. Today, 1989, some 52 percent of the trees are affected by the blight, where needles are lost, and the tree dies. This dieback of forests appears elsewhere; trees in the USA, for example, are exhibiting similar symptoms, though not yet as widespread as in Germany. Scientists blame a combination of factors for the diebacks, with acidifying pollution being only one of many factors at work. As mentioned in the last issue of the *Pulse*, this form of tree death is preceded by a loss of the normal bluish components of both tree and mountain, a blue component which appears identical to the orgone, or life energy charge of the environment. This bluish orgone charge is the "resistance" or "immunity" factor which normally protects the organism from succumbing so easily to multiple environmental assaults.

❖ The amount of true wilderness area on Earth dwindles with each passing year. What little remains has recently been identified in a study by two environmental geographers, J.M. McCloskey and H. Spalding. Using aerial maps from the US Defense Mapping Agency, regions without traces of human habitation were identified, including areas without roads, airports, railroads, pipelines, powerlines, dams, reservoirs, or oil wells. The results: Antarctica 100% wilderness; North America 37.5%, Soviet Union 33.6%, Australia and the southwest Pacific 27.9%, Africa 27.5%, South America 20.8%, Asia 13.6%, and Europe 2.8%. Only 20% of this wilderness land is protected from exploitation, with about half in danger of nearly immediate exploitation or occupation. A large percentage is uninhabitable desert, mountain, or scrubland. The study called for an immediate tripling of the area of protected nature reserves, worldwide. (*Ambio*, V.18,#4)

❖ Recycling efforts around the nation are threatened by an unwillingness of the average citizen to purchase and use recycled products. The USA now ships old newspapers to China, as few of the paper companies in the USA appear interested in making recycled paper products. Part of the reason for this is that the paper mills are owned by the lumber companies, who have no incentives to reduce their tree-cutting activities and use recycled materials. In recent years, many recycled paper de-inking facilities have closed down, due to lack of a market for the ink-free pulp they generate. Many cities also passed laws mandating the collection of old newspapers, largely to reduce the demand for landfill space. However, the cities have done little or nothing about helping to create a market for recycled paper products. New technologies can today create recycled paper products almost indistinguishable from those made out of virgin tree pulp. California may soon pass a law requiring newspapers to use at least 25% recycled paper pulp in the manufacture of newsprint.

❖ The Environmental Defense Fund (EDF) recently published a photograph of Earth from space, showing "giant plumes of smoke from a World Bank project" in the threatened Amazon rainforest. The World Bank pays for construction of "roads to nowhere", going deep into the heart of the Amazon rainforest, allowing the subsequent deforestation of surrounding territory. Giant hydroelectric projects, also funded by the World Bank, flood forested valleys, displace native peoples, and allow boat and barge traffic deep into inland areas that previously were spared from "civilized" exploitation. Slash and burn techniques are used to clear the various areas of their plant and tree cover; the giant plumes of smoke rising from these fires can then be photographed from space. The EDF and other environmental groups have also protested World Bank projects that would have financed roads and harbor facilities into rainforests of Indonesia, for the purposes of tree cutting. As part of these schemes, hundreds of thousands of people from Java would have been forcibly resettled to the Indonesian parts of Borneo and New Guinea. (*Environmental Defense Fund*, 257 Park Ave. S., NY, NY 10010)

❖ The demand in America and Europe for the drug *cocaine* is driving Peruvian farmers to expand coca production into once-protected regions of the Amazon rainforest. Rainforest areas are now being chopped down, or are being sprayed with Agent Orange or paraquat, obtained somehow from sources in the USA. Millions of gallons of acids, acetone, toluene, and other deadly solvents are also dumped into rivers, as by-products of cocaine production. Workers involved in coca production may earn eight times the wage of other farmers. (*San Francisco Chronicle*, 14 August 1989)

❖ Native peoples of the Amazon rainforest have been threatened with forced removal, and possible police actions by various nations that wish to "develop" the rainforest (eg. cut all the trees down, dam up all the rivers, and turn the region into farms and ranches primarily owned by wealthy landowners and multinational corporations). Following a series of protests and legal actions, the following Declaration of Altamira was issued, on 2 February 1989:

"The Indigenous nations of the Xingu, together with relatives from many regions of Brazil and of the world, affirm that it is necessary to respect our Mother Nature. We advise against destroying the forests and the rivers, which are our brothers. We have decided that we do not want the building of dams on the Xingu River, or on other rivers of Amazonia, for they threaten the indigenous nations and riverine populations. For a long time, the white man has offended our way of thinking and the spirit of our ancestors. Our territories are the sacred sites of our people, the dwelling of our creator, which cannot be violated. In the meeting of the indigenous people of Xingu, we have decided to watch over actions of the government to prevent further destruction, to join forces with the National Congress and with the Brazilian peoples, so that together we may protect this important region of the world, our territories."

❖ Read and Weep Department: *"Last year, fires in Brazil produced more than 500 million tons of carbon, 44 million tons of carbon monoxide, 6 million tons of particulates, almost 5 million tons of methane, 2.5 million tons of ozone, and more than one millions tons of nitrogen oxides and other substances that can circulate globally and influence radiation and climate. The amount of Amazon forest burned down is estimate to have doubled in two years leading to increased fears for the future of the ozone layer. For the first time, Brazilian scientists have measured the amount of Amazon forests burned during the annual dry season to make way for cattle pasture and crops. The numbers are terrifying. They found that in 1987 no less than 63,939 square miles went up in smoke... About half the area burned was virgin forest."*

"In 1987 there was so much fire that for weeks all of South America was covered by a dense veil of smoke. In southern Brazil the sky appeared grey with smoke, the sun was weak and dark red and disappeared long before it touched the horizon in the evening. Even in La Paz, Bolivia, at an altitude of 3000 meters, the airport was closed several times for lack of visibility due to the smoke. The extent of the devastation is such that it can only be called a biological holocaust — the likes of which has never happened before in the history of life... This holocaust was triggered to a very significant degree by taxpayers money from the so-called 'First World'. ...agricultural policies of the last 30 years — promoting monoculture cash crops for export — have almost totally destroyed the few existing peasant cultures and devastated the remaining forest. Millions of Brazilians were made landless with the alternatives of either going into the festering slums of large cities or migrating into the remaining wilderness. The World Bank financed the road that opened up [the forest lands]. For the indians, what is happening is genocide... the rubber tappers are also being thrown out of their ancestral lands... More recently, the World Bank promoted another project of incredible devastation,

the Carajas Project [where] the government plans to demolish whole mountains to extract iron and other ores for export... Since there is no coal or petroleum in the area, the pig iron smelters will burn charcoal made in primitive kilns from native forest wood." We note that much of North Africa and the Middle East was deforested during the Iron Age, partly to make charcoal for iron smelting. This is a forgotten lesson from history. (Earth Island Journal, Spring 1989; Rainforest Action Network, 300 Broadway, #28, San Francisco, CA 94133)

❖ Australia's Prime Minister, Bob Hawke, called for a $320 million landcare program for the 1990s, with the planting of a billion trees in Australia. Approximately 50% of Australia's tree cover has been stripped away during the 200 years of settlement by White Europeans. (Climate Alert, Summer 1989)

❖ The editors of the Pulse have vowed to find a way to print the journal on recycled paper, and hopefully this will be done within the coming year. Recycled paper presently costs more, but saves trees, and requires 64 percent less energy to produce. One tree must be cut to produce approximately 100 pounds of printed paper product. As our present small circulation weighs between 100 to 200 pounds, and this is our second issue, we figure we already owe Ma Nature a total of four trees. We have pledged to plant the four trees before the end of this year, and encourage our readers and friends to do likewise. For the record, we got this idea from Earth Island Journal, and join them and several other environmental organizations in leading the way back to a healthy planet. An additional idea: how about a potted Christmas tree which you can plant in the ground after the holidays. Let's see, if you live to be 100, that's 100 trees, times 250 million people...

DESERTIFICATION

❖ Forests are being cut down, and are replaced by grasslands, while grasslands are being assaulted and converted into deserts. Deserts are growing worldwide, with about 70,000 square kilometers of additional desert lands appearing each year. It is a process that has been going on since approximately 4,000 BC, with the onset of human armoring, and the widespread fear and hatred of nature." Dr. James DeMeo has been making this statement repeatedly at talks, and in print, for many years. Recently, confirmation for this view came from a Conference on Implications of Climate Change for Africa, from Col. Christine K. Debrah, Executive Chairman of the Environmental Protection Council of Ghana:

"As we are meeting here today, millions of Africans are suffering from hunger and malnutrition. Vast numbers of people are moving within and across national boundaries in search of food, thereby creating environmental refugees. There is, of course, no accurate measure of the extent of the crisis. But the orders of magnitude are indicative: 150 million people threatened by starvation or malnutrition and an estimated 4 million refugees and returnees and an untold number of displaced persons... Thus, in Africa, dry weather conditions have been experienced leading to bush-fires and further degradation of the land. There is, to echo the Executive Director of the United Nations Environment Programme in Africa, creeping 'Savannization of the forest, Sahelization of the savanna, and desertification of the Sahel'." (Climate Alert, Summer, 1989)

DEATH OF THE OCEANS

❖ New distressing information on the deterioration of life in our oceans:

- Algae blooms in the oceans are usually related to dumping of raw sewage, agricultural runoff, and other pollutants, and have killed fish in areas as large as 400 square miles. Globally, the phenomenon of algae blooms ("red tides") is on the increase, and cannot be attributed to "nature". They kill fish and marine mammals, render shellfish toxic to eat, and make beaches stink.

- Almost 50% of the Baltic Sea's bottom waters are oxygenless.

- Algae blooms in the North Sea have recently increased by 400%. One toxic algae bloom in Skagerrak, connecting the North Sea with the Baltic Sea, killed all marine life to a depth of 50 feet.

- At least six major toxic blooms of red tide have occurred along the East Coast of the USA since 1972, in regions where they were once quite rare.

- One form of algae common only in the Gulf of Mexico recently struck beaches near Cape Hatteras, North Carolina, inflicting a $25 million loss on the area's fishing and tourism; 3,000 dolphins were also killed, many washing up on public beaches.

- Guatemala and the Philippines have been reporting deaths from contaminated shellfish, an ailment completely unknown in that region prior to the 1980s.

- Nearly 300,000 people contracted hepatitis-A within 3 months in Shanghai, China, due to contaminated clams; 47 people died.

- At any one time, 33% of all US shellfishing beds are closed due to pollution contamination.
(World Watch, July/Aug. 1989)

❖ The supertanker Exxon Valdez dumped thousands of barrels of crude oil in Alaska's Prince William Sound, on March 24, quickly coating over 730 miles of wilderness coastline with stinking goo. In some places, the oil slop was three feet deep. Little of the oil has been recovered, though a few of the more important and visible beaches have been cleaned. In many areas, the oil simply fell to the bottom of the Sound, where it remains, waiting to be transported by heavy winter waves. (World Watch, July/Aug. 1989)

❖ A study by the US National Research Council indicated that 21 million barrels of oil enter the sea from small and steady sources, worldwide. Street runoff, ships flushing their tanks, and effluent from industrial facilities are the sources of this oil, which exceeds the 600,000 barrels that are accidentally spilled each year. As little as 1 part of oil per 10 million parts of seawater has serious effects upon the growth and reproduction of plankton, crustaceans, and fish. (World Watch, July/Aug., 1989)

❖ Each year, 30,000 northern fur seals die after becoming tangled in plastic bags, lost fishing nets, or other forms of plastic junk. An estimated 500,000 plastic containers are dumped into the sea each *day* by merchant ships, and hundreds of miles of fishing gear, including whole nets, are accidentally lost. (*World Watch*, July/Aug. 1989)

❖ *"Every night, more than 700 fishing boats equipped with 20-mile long drift nets sweep an area of the northern Pacific the size of Ohio."* With large fleets of fishing boats, and factory processing ships that work for months at a stretch out at sea, the oceans are being harvested beyond their sustainable yields. *"The oceans could soon lose their vitality if people aren't encouraged to become stewards of a shared resource, rather than plunderers of a common frontier".* (*WorldWatch*, July/Aug. 1989)

❖ Chinook and coho salmon caught in coastal hatchery runs in northern Washington state are routinely exhibiting the symptoms of viral hemorrhagic septicemia (VHS). This disorder is worse for fish than AIDS is for humans, according to experts on the subject. Other regions have instituted a ban on importation of fish eggs from Washington. Additionally troubling is the fact that environmental toxins and immune depletion play a role in determining the susceptibility of an organism to VHS. In this context, it is significant that tons of Canadian and USA chemical warfare weapons, and toxic wastes from industrial companies have been dumped into the ocean off Neah Bay, in the Strait of Juan de Fuca at the Canadian/USA border. The region is marked "restricted" by the US Navy, and is very close to the affected hatcheries. Apparently the fish cannot read the warning signs. (*Acres USA*, July 1989)

❖ Sea levels around the world have already risen by from 10 to 20 centimeters over the last 100 years, with a projected rise of around 0.24 centimeters per year. A sea level rise of from .5 to 1 meter may occur over the next 100 years, as a result of glacial melting and global warming. In Miami, levels have risen observably since the 1940s, as documented by changes in

barnacle levels on sea walls. Each coastal town or facility will have its own specific difficulties to content with, given that some coastal areas are either rising or subsiding in relation to this sea level change. A one meter rise is quite significant, however, and will endanger and inundate many coastal areas. (*Science News*, 10 June 1989)

THE DESTRUCTION OF ECOSYSTEMS

❖ A recent National Research Council report has indicated that agricultural productivity in the worlds "breadbasket" regions have nearly reached their yield limits, and may soon begin to show decreased yields. The anticipated declines are due to mismanagement of soils, from failures to use time-tested methods of contour plowing, strip cropping and crop rotation, from plowing of marginal and steeply-sloped lands, and from unecological methods of crop production, to include widespread use of salt-based fertilizers, harsh pesticides and herbicides, and failures to reintroduce organic material into the soil. Soil erosion is anticipated to increase worldwide, as populations continue to soar, and as marginal lands are increasingly brought into production. One particularly telling example of this process can be seen in Haiti, where centuries of deforestation and environmentally-unsound farming practices have exposed soils to tropical rains. The island is today surrounded by a brownish "bathtub ring" of eroded soil, which extends for miles out into the normally blue ocean water, and can be seen from the air. On another island, Madagascar, deforestation is likewise severe, and much of the topsoil is now washing out to sea. In the overpopulated regions of northern India and Nepal, deforestation is likewise severe, such that a near total ban on woodcutting may go into effect, in spite of the devastating consequences to local peoples. Soil erosion from these deforested areas has badly silted the Ganges River. Silting in the downstream areas of Bangladesh has resulted in an increasing frequency of floods. (*Science News*, 15 July 1989)

❖ An international ban on trade in ivory is in the works, following revelations that the African elephant herd may be in danger of extinction. In recent years,

bans on hunting of elephants have caused the price of ivory to soar on world markets, fueling an increase in poaching within game preserves. Park rangers in Kenya and Tanzania, for example, now carry automatic weapons, and a "shoot to kill" policy exists against poachers, who have murdered many unarmed park rangers in recent years. Japan, Hong Kong, the USA and the EEC nations have voted to severely limit or ban ivory imports, but it remains to be proven that this will be effective. Some ecologists feel that this step is too little, too late, and that the elephant is probably doomed.

❖ A minimum of one fourth of the Earth's total number of animal, plant, and microbe species will soon be extinct unless measures are taken to protect them, says a recent National Science Foundation study. While 1.4 million species have been named and catalogued, an estimated 5 to 80 million others remain unstudied, primarily in the threatened lands of the tropics and subtropics.

❖ *"Each year we see harvests are smaller, the deserts are larger, the topsoil is getting thinner, the ozone layer is depleted, the level of greenhouse gases in the atmosphere is building, plant and animal species are getting fewer..."* Lester R. Brown, Worldwatch Institute.

❖ *"A world where industrial production has sunk to zero. Where population has suffered a catastrophic decline. Where the air, sea, and land are polluted beyond redemption. Where civilization is a distant memory... This is the world that the computer forecasts. What is even more alarming, the collapse will not come gradually, but with an awesome suddenness, with no way of stopping it."* (From the censured 1972 Report of the Club of Rome, *The Limits to Growth*.)

❖ China has been losing more than 1,000,000 acres of farmland each year, for the past three decades. Some of this is due to creeping desertification, but a major portion is due to spreading urbanization, or loss of farmlands for living space in an overcrowded landscape. China has only about half the arable land as does the

USA, but nearly four times the population. (*WorldWatch*, July/August 1989)

❖ *Environmental threats with the potential to erode the habitability of the planet are forcing humanity to consider national security in far broader terms than that guaranteed solely by force of arms".*

So states a recent Worldwatch Paper on *National Security: The Economic and Environmental Dimensions*, by Michael Renner. World-wide, the inflation-adjusted expenditures on military matters has increased from 200 billion dollars in 1950 to nearly 1 trillion dollars in 1987. Three nations, the Soviet Union, mainland China, and the United States together account for over half of the world's soldiers and workers involved in military employment, some 24 million individuals (mostly men) out of a global total of at least 45 million. The Soviet Union, in 1985, spent 12.5% of its GNP on military matters, compared to 6.7% and 6.6% for China and the United States, respectively. Approximately 2 days of global military spending would fund the yearly cost of a proposed United Nations program to halt desert spreading around the world. (*Worldwatch Institute*, 1776 Mass. Ave., NW, Washington, DC 20036)

ENDANGERED ENVIRONMENTALISTS

❖ Environmentalists who take active and effective measures to protect wilderness areas or wildlife are increasingly threatened with death. Some examples: Chico Mendes, an officer for a Brazilian rubber tappers union, was shot down for his work protecting rainforest regions from land developers. Six previous attempts against his life were never investigated by the Brazilian government. Barbara D'Achille, one of Peru's better-known environmental journalists, was murdered (stoned to death) by Sendero Luminoso guerrillas, for having written about the environmental destruction associated with coca cultivation. Joy and George Adamson, who wrote the popular book *Born Free,* about the lioness *Elsa*, were both murdered while working to protect the African lion, which is heavily poached. Joy was stabbed to death in 1980, and George was machine-gunned most recently. Diane Fossey was axed to death

by poachers for her work protecting the African gorilla. An environmental photographer was murdered by French terrorists, who bombed the Greenpeace ship *Rainbow Warrior*, which had documented the great damage from French nuclear testing in the Pacific. Members of environmental groups in the USA have recently been infiltrated by the FBI, and targeted for arrest and possible lengthy jail terms for taking symbolic actions against either irresponsible forest clear-cutting practices in the Pacific Northwest, or against nuclear power facilities. Confrontations between groups of environmentalists and native peoples on the one side, and governments, bankers and developers on the other, are heating up all around the world. In general, the more effective are the legal or direct actions taken by environmentalists, the more personal danger they face.

❖ *"Ours is the first generation faced with decisions that will determine whether the Earth our children inherit is habitable."*
— Lester R. Brown

POSITIVE SIGNS

❖ Earth Day, 1970, will be repeated next year, by an organization with the appropriate name: **Earth Day 1990.** To obtain more information, write to: Earth Day 1990, PO Box AA, Stanford University, Stanford, CA 94305.

❖ *A Green City Program, for San Francisco Bay Area Cities and Towns* has recently been published by the Planet Drum Foundation. The book outlines many constructive things that can be done to prevent the Bay Area from deterioration into the typical urban mess so characteristic of other large metropolitan areas. A list of local citizen's working groups is provided, for those who want to become actively involved. However, the general suggestions and plans in the book could be applied to any other city with great benefit. Planet Drum is developing concrete programs for halting the deteriorating social, economic, and environmental situations of the other bioregions in the USA. (Contact: *Planet Drum Foundation*, PO Box 31251, San Francisco, CA 94131, Shasta Bioregion, USA)

❖ New laws from Berkeley to Minneapolis to New York to Houston will soon ban the use of non-recyclable, non-degradable food packaging, at both fast food outlets, and in supermarkets. Polystyrene and polyvinylchloride products, as used in plastic tableware, styrafoam cups and packaging, plastic milk jugs and pop bottles, and clear plastic packaging for department and hardware store items, will be the first items to vanish. This is a necessary and growing trend, which will help eliminate toxic chemical by-products, and diminish the piles of garbage in our landfills.

❖ Solar powered automobiles may be in your future. Solar vehicles that can go up to 40 mph, with a cost below $5,000, are already feasible, and projections exist for 60 mph versions at about three times this cost. Sound unbelievable? Just remember the first airplane flight, which lasted less than 30 seconds, going about 10 feet off the ground, at about the speed of a running horse. And that was only about 80 years ago. We have recently seen a solar powered aircraft fly to several thousand feet. Now, what about an aircraft powered by an orgone motor?

❖ *Pulse of the Planet* wants your comments, suggestions, and newspaper clippings on materials you believe would be helpful or important for our readers. Please send them to the *Pulse,* PO Box 1395, El Cerrito, CA 94530.

What you can do:
Go to your Public Library, and obtain a list of local environmental groups. Offer your assistance. Participate in whatever local recycling programs are available in your region. Educate yourself about renewable energy systems, and reduce your own personal impact upon the planet.

❖ Wilhelm Reich's discoveries on the biologically exciting, and potentially therapeutic nature of the oranur effect, which could drive latent health problems to the surface (or, alternatively, cause flare-ups that were potentially deadly), appears to have been partly rediscovered by scientists researching the phenomenon of "hormesis". Hormesis is a word coined in 1940, possibly to avoid using the terminology of homeopathy, to describe the beneficial effects of very low doses of chemical agents which are harmful in larger doses. In this case, however, the researchers were investigating the effects of low level ionizing radiation, which have been observed to have a "stimulating" influence. Recent experiments suggest that mice exposed to such low dose radiations actually live longer. These observations bring to mind some of the older radiation therapies, dating back to the 1800s, where people drank radium-affected water, or sat in abandoned uranium mines, claiming a therapeutic effect. Skeptics of this effect point to conventional radiation theory, which cannot account for it, and to other studies on the mortality and morbidity resulting from low-level radiation exposure.

Unfortunately, the promoters of nuclear power plants are most enthusiastic about the finding of hormesis, as a means to justify exposing the general population to additional low level radiation. ("It's good for you!") Whatever, these findings suggest that both hormesis and homeopathic effects may in reality be *oranur* effects, which develop in organisms exposed to life-negating toxins, but which cannot be "diluted out" of a homeopathic preparation. (*Science*, 11 August 1989)

❖ *Time* magazine (18 September) recently ran a photo article on mysterious swirling circle patterns in British wheatfields. They appear to be recent landing sites for UFOs, which possibly came to Earth for a few minutes or hours, and then flew away, leaving behind the circular imprint pattern in the cropland. Two new books on the subject have appeared: *The Circles Effect and its Mysteries*, by T. Meaden, and *Circular Evidence*, by P. Delgado and C. Andrews. We have no observed reports on this to make, but would consider publishing any materials

on this, or other aspects of the UFO question that our readers would care to submit.

❖ More information on the "Blue Fuzzies" seen by astronomers. The fuzzy blue dots are extremely faint and require hours of exposure time to register on photographic plates. Even so, no details can be made out. There are also uncountable numbers of them, as many as 25,000 in the area the size of the full moon. The astronomers think they are very far and distant galaxies, out at the edge of the universe. But anyone with a relaxed eye who looks out into the open blue sky on a clear, crisp day will likely get an idea of one phenomena which the astronomers may be photographing upon their plates, namely the *orgone units*, or *pin-points* of light that can be seen with the unaided eye. Unless and until the astronomers can demonstrate that these "blue fuzzies" appear with the same pattern on separate photographic plates, taken at different times, or until they can provide detailed photos, revealing some kind of known galactic structure to them, an orgonomic interpretation of the effect is reasonable. (*Pulse of the Planet*, Spring 1989)

❖ The German scientist Dr. Fritz-Albert Popp recently confirmed that a very weak light is given off by living tissue ("Ultraweak photons"), and can be detected with sensitive photomultiplier tubes. This light given off by cells, which Popp calls "biological luminescence", can vary somewhat depending upon the state of health of a given cell. Popp has argued, and presented supporting evidence, to show that energy emitted from malignant cells in one culture dish can trigger healthy cells in a nearby culture dish to likewise become malignant. Japanese researchers are also investigating methods for measuring the "Light from Living Cells". Funded by 5 billion yen (about $5 million), a research team led by Dr. Humio Inaba of Toboku University will investigate the subject of "Biophotonics". They foresee new diagnostic medical instruments, as well as basic biological discoveries to arise from the work. Like Popp, Inaba has found that cell division, blood plasma, and

other "excited" tissues give off the greatest amount of light. These findings confirm a basic energetic parameter at work regarding the structuring and regulation of cells, and thereby confirm portions of Reich's work on the cancer biopathy, and the luminous orgone energy field of cells and organisms. (*Int. Journal of Fusion Energy*, April 1985, October 1985; *New Scientist*, 27 May 1989). Also see the "Letters" section of this issue of the *Pulse* for discussion on methods for photographing weak light emission from living tissue, without using electricity or electronic instruments.

HEALTH NOTES

❖ A recent book by Dr. Joseph Sacco (Morphine, Ice Cream, Tears, 1989) documents the abusive treatment new medical interns and residents are subjected to by their superiors. Sacco tells how "...his residency in a city hospital nearly ground his high principles into dust. Bone-weary from endless hours of duty, aghast at the treatment of helpless patients... he records the greed and arrogance of medical 'stars', warns about turning one's fate over to a high-tech medical community that, in its zeal to attack the disease, often kills or maims the patient." For the young doctors, the scenario sounds much like military boot camp practices, which are designed to produce mindless obedience, while reducing one's sense of emotional empathy to persons outside of the group, namely the hospital patients.

❖ For those wishing to document the absence of effectiveness of classical treatments for cancer, documentation can be found in the study of J. C. Bailar and E. M. Smith ("Progress Against Cancer?", *New England Journal of Medicine*, 314:1226-1232, May 8, 1986) and in the report by H. Bush (*Science 84*, September 1984). Both of these studies demonstrate that survival rates for cancer patients subjected to the conventional treatments of surgery, radiation, and chemotherapy is no better today than in the 1950s.

❖ The Congressional Record Appendix of August 3, 1953 contains an amazing report by Benedict F. Fitzgerald, Jr., Special Counsel to the Senate Committee on Interstate and Foreign Commerce. Fitzgerald investigated the operations of over 30 foundations, hospitals, clinics and government organizations specializing in cancer (including the American Medical Association and the National Cancer Institute), and *confirmed* that there was a conspiracy to suppress certain innovative cancer treatments. Copies of this testimony, and other informative materials on alternative cancer treatments, can be obtained from: *Project Cure*, Center for Alternative Cancer Research, 1101 Conn. Ave., NW, #403, Washington, DC 20036.

❖ The following news report recently came to our attention: **Reagan Seeking Health Alternatives from Foreign Cancer MD?** *"While the government supported suppression of freedom of choice in cancer therapy continues unabated, the President of the United States is apparently seeking one of the suppressed alternatives. According to reports from European newspapers and magazines, the White House has been quietly in touch with Dr. Hans Nieper, purveyor of 'forbidden' cancer remedies. The German publication* Neue Revue, *in an article headlined 'A Call from the White House: Save Reagan' said that Nieper... has been in contact with the president and his physicians..."* (*Spotlight*, 6/13 January, 1986)

❖ Following a series of recent police actions against alternative and holistic health practitioners, new medical treatments and health food stores, involving sweeping "gestapo type" tactics, a new Coalition for Alternatives in Nutrition and Healthcare (CANAH) has been formed to institute a Healthcare Rights Amendment. The amendment reads as follows:

"Healthcare Rights Amendment: Section 1. The Congress shall make no law which restricts any individual's right to choose and to practice the type of healthcare they shall elect for themselves or their children for the prevention or treatment of any disease, injury, illness or ailment of the body or the mind. Section

2. The Congress shall have the power to enforce, by appropriate legislation, the provisions of this article. Section 3. This amendment shall take effect immediately after the date of ratification."

Individuals interested in learning more about the abuses of state power by medical organizations should contact: CANAH, PO Box B-12, Richlandtown, PA 18955.

❖ Several Food and Drug Administration scientists were recently indicted for accepting bribes from drug companies. According to one source: *"The FDA has been caught with its hands in the cookie jar. It has been general knowledge that the FDA was on the 'take' for 30 years and that it was 'taking' from pharmaceutical companies without regard for the public welfare. The fact that drugs with fatal side effects were approved by the FDA, and effective drugs that had no side effects were dumped was a clear sign that there were unseen forces involved other than 'truth in science'".*

The FDA, which quickly impounded tons of Chilean grapes after finding cyanide in three grapes, has indicated that they will not be pulling any of the questionable drugs off the market. They have also continued to do "business as usual" with the offending companies that bribed the FDA scientists. (*Donsbach Report*, July 1989)

❖ UNICEF reports a 25% to 50% reduction in per-capita health and education expenditures among the 37 poorest nations on Earth. A similar trend prevails among nations that are not among the poorest of the poor. The total number of persons living below the poverty line is greater in the early 1980s than in the early 1970s. These depressing trends are the result of growing populations, declining agricultural and economic productivity, declining natural resources, decaying ecosystems, and growing economic debt. Malnutrition also increased; daily caloric intake among 21 of 35 low-income nations decreased in the 1980s to levels lower than in 1965. (*WorldWatch*, July/August 1989)

❖ McDonald's hamburger restaurants take in about 10% of all money spent by Americans on restaurant food, and they have branches in major cities all around the world. These kinds of fast food restaurants have often been attacked, for selling unhealthy "junk food" to children. But it seems that we have been misinformed, for McDonald's restaurants now sells only "nutritious food". On several poster-placemats, with propaganda designed for ignorant adults and children, the restaurant tallies the vitamin contents of a constipating concoction of cheeseburgers, chicken nuggets, french fries, and soda pop. Nowhere is it indicated that these foods are saturated in fat, sugar and salt, and have a very low fibre content. On one place mat, the "Real Seal" of the American Dairy Association is boldly displayed in front of a chocolate milk shake and hot fudge sundae.

"Instead of the empty calories you get in junk food, our meals provide the vitamins and minerals so essential to your health" states one of their posters, which appears designed for schools. And yes, some schools have invited McDonald's or other junk food restaurants to serve up burgers and fries to school kids, insuring that their nutritional needs would get the same inadequate attention as their emotional and educational needs. Worse, we now hear, from National Public Radio, that the Los Angeles Children's Hospital has signed a 20 year contract to provide McDonald's food right in the hospital, for the sick children and their parents! The Los Angeles Children's Hospital is the 11th hospital in the nation to take this dangerous step. However, we are informed that McDonald's does offer a charity vacation retreat for children who have been given colostomies to "cure" their intestinal problems

The term "sex-economy" was developed by Wilhelm Reich in the 1930's to describe the social and biological influences upon the expression of the life energy in an organism. In particular, he observed that societies take a keen interest in regulating the flow and distribution of biological energy as expressed in human behavior, sexuality, and emotion. Laws and belief systems are developed to define the fundamental, energetic relationships between men and women, children and parents, and people in relation to the natural world. These laws and beliefs shape the expressive nature of biological life. Many laws and practices attempt to repress and kill the natural expression of the life energy; and there are some that bravely protect it. In this section, we will report on both.

❖ A United Nations working group on slavery reports that 1 million children a year are forced into prostitution. The underage sex racket thrives in areas where government authorities are unwilling to stop it, as to do so would hurt their "tourist" trade. (*San Francisco Chronicle*, 10 August 1989 For those who have the stomach to read about it, the classic work by Kathleen Barry (*Female Sexual Slavery*, 1981) spells out the dirty traffic in women and children as it exists around the world. A recent specific example comes from Nepal, where an organized traffic in the sale of young girls for prostitution in India was uncovered. Social workers in the area state that around 3000 girls between the ages of 9 and 20 are lured from rural Nepal each year, on the promise of jobs or husbands. They are instead sold to brothels in distant cities and kept as virtual prisoners. More stringent laws have been introduced to curb the abusive practices, but the high prices gained from the traffic in girls and women, the very low social status and perceived value of females, and the grinding poverty of the overpopulated regions, work to thwart such reformist efforts. (*WIN News*, Winter 1989)

❖ There are approximately 40 million homeless children around the world, living in the poorest parts of the world's largest cities. Latin America faces the biggest problem with these children. Roughly 30% have been abandoned by their families, or have run away, and are entirely on their own. Considered hooligans or delinquents, it is no surprise that many turn to stealing, prostitution, and other forms of street hustling as a way of surviving. In Manila, the Philippines, as many as 30% of the citie's 15,000 street children are involved in prostitution. Rather than being helped, they are usually greeted with punishment and police harassment. The problem is most severe in those nations where the means for preventing unwanted pregnancy—contraception, abortion, and sexual hygiene education —are either illegal or unavailable. (*WorldWatch*, July/August 1989)

❖ Researchers studying the "cycle of violence" theory, that abused and neglected children grow up to be abusive parents and violent criminals, are failing to fully confirm the prediction. Some studies show that only slightly more abused children grow up to be violent, than non-abused children. (*Science News*, 22 July 1989) Unfortunately, none of the studies in question attempt to include the full array of sex-economic factors identified by Reich as causing violent behavior. Primary on this list is the sexual life of the child and adult. Reich demonstrated that if the child and older adolescent could establish a healthy love life, that the affect of earlier abuse could be mitigated somewhat. Also, chronic sexual frustration (absence of gratification, and not necessarily absence of "sex") would itself lead to violent behavior, even among those who had *not* been violently abused. Indeed, Reich identified a special *impulsive character* who was exceptionally violent, based upon the fact that they had been given great freedom in youth, except for sexual freedom; this caused a great increase of dammed-up sexual tension and frustration in the individual, but without the usual characterological means for holding such anger in check. Outbursts of violence would occur in such people. These abusive and sexual factors were reviewed by both Prescott and DeMeo in separate cross-cultural studies: there is a clear capability of predicting the levels of violence in a given society, but only when the factors of sexual repression are also studied. Child abuse by itself is not as powerful a predictor of violent behavior in adulthood as is sex-repression. While our culture is now coming to grips with the problems of child abuse, we have yet to deal openly with the question of adolescent sexuality, except to suppress that sexuality, or confound it with guilt, anxiety, and fear. These latter factors clearly lead to a reduction of capacity for sexual tenderness and gratification, thereby fostering an increased sexual frustration, internal tension, and violent discharge.

❖ The examples of environmental abuse, deforestation, and soil erosion given in the Environmental Notes, are underlain by one major set of sex-economic factors: an explosive population growth which leads to increased competition for resources that simply cannot provide for everyone on a sustained basis. Such explosive growth of population is underlain by antisexual and anti-contraceptive policies of various governments and religious organizations, world-wide. In Latin America, the primary culprit on this list is the Catholic Church, which has for centuries issued powerful antisexual propaganda. For example, the overcrowded island of Haiti reached its natural sustainable population limit years ago, but populations continue to expand. Tree-cutting and land clearing, for the purposes of agriculture and firewood, has nearly denuded the entire island, including many steeply-sloped areas. Bare soils have washed into the sea. Agricultural productivity consequently has declined, requiring massive foreign food aid to prevent starvation. As the population of Haiti continues growing, with no place to go, those who can do so try to leave by boat to other Caribbean nations, just as people from overcrowded India and China are migrating to almost any other nation that will take them in. In Brazil, the Pope was recently heard to damn women who use *contraception* (and not just abortion) as "murderers", implying that life begins even *prior to conception*, and that a man's sperm was worth even more than the life of a woman. Wife-murder is currently tolerated under Brazilian law, assuming that the man can prove that his wife "dishonors" him. Massive and widespread infanticide of living babies, generally all those born after the 6th child, continues in many parts of Catholic Latin America, where the infants are often tied to a bedpost and starved to near death. At the last minute, parents rush the unconscious, famished baby to the doctor, pleading for help. Such are some of the devastating social factors underlying the environmental devastation in those regions, where the Catholic Church continues to loudly beat the anti-contraceptive drum. From this, it is no wonder that the lives of the indigenous natives, who are not Catholic, and who often have a healthy attitude regarding sexual pleasure, are ignored in the rush to develop the rainforests.

❖ Reports from India and Pakistan (short wave radio) continue to suggest that there is a vicious trade in human organs for the hospital surgical transplant market. According to the reports, from BBC and elsewhere, poor people can sell a kidney or other body part to a sadistic doctor or corrupt hospital, to be transplanted into a rich person who is too impatient to wait for a donated organ from a fresh cadaver, or too brainwashed to consider alternative treatments. Police in India recently rescued a dozen women and children who had been sold by their husbands or parents for their organs; these poor souls would have been butchered and dissected, their organs then passing directly into the global, underground organ transplant marketplace. Human life is cheap in areas where centuries of antisexual attitudes have led to anti-contraceptive policies, and explosive population growth rates. Both Muslim and Hindu marriages in the region are almost entirely arranged, and open expression of heterosexual romantic feeling by young people can lead to murder of the offending child by the parents, for the sake of "family honor". Even Mother Theresa sternly refused to consider birth control as a means to reduce the abandonment of infants left on the stoops of temples and government buildings. The situation in India is so bad that, only a few years ago, poor urban women in India protested for a return of the ritual burning of widows, as society had no place for a widow but the street. With so many children, and so few natural resources in the region, families are manacled to a cycle of poverty. Low caste families of crowded urban areas have been known to purposefully cripple or amputate the arms or legs of their own living children, in order to make them "better beggars", more capable of eliciting sympathy and token money from passers-by. Such is the hidden face of Eastern anti-sexual, anti-contraceptive mysticism.

❖ *"According to the United Nations, women do two thirds of the world's work for 10% of the income, and own only 1% of the assets"* — Mildred Gorden, Member of the British Parliament, speaking before the House of Commons, April 1989.

❖ The above quote by Mildred Gorden is a major underlying factor in the desperate condition of many of the families in the USA where the father is absent, and mother is left to provide for the children by herself. In such cases, it is usual that the family bonds were fabricated in order to meet the social demands surrounding an unwanted and unplanned pregnancy. This is most typical in those cultural groups where good sex education, and access to contraceptives are absent. While almost every corner drug store carries contraceptives, the current costs are often beyond the reach of teenagers and poor women, and it is a fact that the typical "macho" male leaves contraception entirely up to the female. Indeed, "knocking up" a girl is considered to be proof-positive of a virile sexuality among male teens living in some ghetto poverty areas. This attitude developed on the streets, in a vacuum left by sex-education courses that rarely discussed sexual feeling in a positive manner, and usually attempted (without success) to promote "abstinence" among sexually preoccupied and energized adolescents. Fully one fifth of the nations children now grow up in poverty circumstances, with an increase in the number of children having children. The connection between unplanned pregnancy, contraception, family size and poverty, along with the economic advantage of males over females, and the hierarchical structure of our economic system (where a tiny percentage of people own and control the greater portion of the property, assets, and wealth), must be confronted if our present downward spiral is to be halted.

❖ The Third Edition of *The Hosken Report on Genital and Sexual Mutilation of Females*, by Fran Hosken, has recently come across our desk. It is undoubtedly the best source of information on the current status of the female mutilations. Her other publication, *WIN News*, is a periodical that is one of the best sources of information on sex-economic factors around the world that is to be found, and we highly recommend it. She has been a tireless fighter for the rights of women and children. Her publications may be ordered from: *Women's International Network News*, ~~187 Grant St., Lexington, MA 02173.~~ An article by Ms. Hosken will appear in a future issue of the *Pulse*.

❖ Dr. James Prescott has recently taken over editorship of the *Truth Seeker*, a "Freethinker's Publication" that has been in print since 1873. The most recent issues have dealt with subjects that rarely get much press exposure. The March/April issue dealt with the subject of "GAIA: A New Look at Life on Earth", the May/June issue with "The Liberty of Man, Woman and Child", and the July/August issue with "Crimes of Genital Mutilation". This latter issue included articles by Prescott, DeMeo, Marilyn Milos (Director of NOCIRC) and Fran Hosken (Director of WIN News: see note above), which were presented earlier this year at the First International Symposium on Circumcision. A future issue of the *Pulse* will present an article on "Body Pleasure and the Origins of Violence" by Dr. Prescott.

Two items below are reprinted from the *Truth Seeker*; "Sixteen Principles for Personal, Family and Global Peace", and the "Declaration of the First International Symposium on Circumcision". (*Truth Seeker*, ~~PO Box 2832, San Diego, CA 92112~~)

SIXTEEN PRINCIPLES FOR PERSONAL, FAMILY, AND GLOBAL PEACE

By James Prescott, Ph.D.

1. Every pregnancy should be a wanted pregnancy. Every child should be a wanted child.

2. Every pregnancy should be free from alcohol, drugs, tobacco and other harmful agents of stress.

3. Every pregnancy should have proper nutrition and health.

4. Every normal birth should be without drugs.

5. Every birth should be a loving event with family and friends.

6. Every baby should be breastfed for two years or longer and be given a loving massage every day.

7. Every baby should be carried on the body of its mother, father, or caretaker as much as possible.

8. No baby, child or person should be subjected to any form of genital mutilation for reasons of religious belief or social custom.

9. No baby or child should be hit, spanked or humiliated.

10. No baby or child should be left to cry itself to sleep.

11. The personal dignity of every baby and child should always be respected and affirmed.

12. The emerging sexuality of every child and adolescent should always be respected and honored.

13. The right of self-determination in the sexual expression of affection and love is a basic human right of all persons.

14. Every human being should receive a loving massage every day.

15. Sexual affection and sexual love are essential wellsprings for human peace and harmony.

16. The home and family is the cradle of alienation and violence, or the cradle of love and universal peace.

(From: *Truth Seeker,*
March/April 1989, p.33)

DECLARATION OF THE FIRST INTERNATIONAL SYMPOSIUM ON CIRCUMCISION

We recognize the inherent right of all human beings to an intact body. Without religious or racial prejudice, we affirm this basic right.

We recognize the foreskin, clitoris, and labia are normal, functional body parts.

Parents and/or guardians do not have the right to consent to the surgical removal or modification of their children's normal genitalia.

Physicians and other health-care providers have a responsibility to refuse to remove or mutilate normal body parts.

The only persons who may consent to medically unnecessary procedures upon themselves are the individuals who have reached the age of consent (adulthood) and then only after being fully informed about the risks and benefits of the procedure.

We categorically state that circumcision has unrecognized victims.

In view of the serious physical and psychological consequences that we have witnessed in victims of circumcision, we hereby oppose the performance of a single additional unnecessary foreskin, clitoral, or labial amputation procedure.

We oppose any further studies which involve the performance of the circumcision procedure upon unconsenting minors. We support any further studies which involve identification of the effects of circumcision.

Physicians and other health-care providers do have a responsibility to teach hygiene and the care of normal body parts and explain their normal anatomical and physiological development and function throughout life.

We place the medical community on notice that it is being held accountable for misconstruing the scientific data base available on human circumcision in the world today.

Physicians who practice routine circumcisions are violating the first maxim of medical practice, "PRIMUM NON NOCERE", *First, Do No Harm*, and anyone practicing genital mutilations is violating *Article V of the United Nations Universal Declaration of Human Rights:* "NO ONE SHALL BE SUBJECTED TO TORTURE OR TO CRUEL, INHUMANE OR DEGRADING TREATMENT..."

Adopted by the General Assembly, March 1-3, Anaheim, CA.

❖ Over 1,000,000 teen pregnancies—almost all unplanned and unwanted—occur in the USA each year. These teenage pregnancy rates are twice those in Britain, three times those in Sweden, and ten times that of Holland. The low teen pregnancy rates overseas has been attributed to the existence of special clinics for young people, near to the schools or actually within them, in which sexual hygiene and education, and contraceptives can be easily obtained. Unlike the USA, they do not require pelvic exams of teen girls prior to dispensing contraceptives, nor do they emphasize "abstinence" over contraception. The goal in the overseas clinics is to prevent pregnancy, not intercourse. The relationship between girls and doctors or nurses is also protected with strict confidentiality. Many girls would otherwise be at risk of physical harm from abusive parents if their sexual activity was revealed — the young girls know this and often will not seek out help from nurses or doctors whom they know will "tell on them" to their parents. Aside from the staggering human toll which comes from unplanned pregnancy, some $19 billion dollars is spent each year in the USA on welfare and other assistance to unwed mothers.

❖ Virtually all controlled studies contrasting the outcomes of birth according to the training and skills of the birth practitioner have demonstrated that *the more classical medical training the practitioner has, the worse is the outcome for both mother and child.* This startling conclusion, which runs counter to prevailing pro-medical propaganda, was made by Dr. David Stewart, an epidemiologist who reviewed all published medical studies which contrasted birth outcomes by the training of the birth attendant. Steward demonstrated that empirically trained lay midwives had the best outcomes of all studied groups, even in high risk situations. Lay midwives had better outcomes than nurse midwives, but both lay and nurse midwives had better outcomes than general practitioner medical doctors. The worse outcomes, for both mother and baby, was found to occur when childbirth was handled by obstetrician-gynecologists. Stewart attributes his finding to the fact that doctors, and obstetricians in particular, are trained in surgery and drugs, and hence apply their training in situa-

tions where they are not called for. Furthermore, he notes the absence of study of basic emotional and nutritional factors by doctors (and even nurse-midwives), along with a distrust of basic life functions, such as childbirth. The medical profession labeled childbirth a pathology requiring medical intervention centuries ago, a step that was never supported by any scientific evidence. Normal and routine birth was therefore made dangerous and risky by virtue of medical meddling, and basic ignorance of doctors about how natural birth actually occurs. This latter observation is clearly supported by the current epidemic of caesarean births, which have soared to around 25% of all American births (and often reach nearly 100% in many teaching hospitals). Stewart's book, titled *The Five Standards of Safe Childbirth* (NAPSAC, PO Box 267, Marble Hill, MO 63764, USA), unmasks the lie that "doctors know best" about childbirth, and, indeed, is a damning indictment of a horrible toll in human misery and death which occurs at the hands of American medical doctors. Based upon the best-available scientific evidence, Stewart lists the five standards for safe childbirth: Good Nutrition, Skillful Midwifery, Natural Childbirth, Home Birth, and Breastfeeding.

❖ Rosalie Tarpening, a 62-year old California lay midwife, was recently found guilty of "second-degree murder" in the January 1988 stillbirth of a baby girl. Tarpening's record of decades of uneventful and safe deliveries was not given weight in the trial, and she now sits in jail awaiting sentencing. By contrast, we point to the fact that no hospital obstetrician was ever put on trial for the death of a baby. When a baby dies in a midwife's care, it is "murder"; when it dies in the hospital, it is "God calling the baby home", or "we did our very best to save it". Even though many parents have sued doctors and hospitals over substandard and dangerous procedures, and outrageous medical incompetence, no prosecutor has ever pressed charges of any kind against a hospital doctor, even though the best scientific evidence shows that midwives delivering at home have a safer track record than doctors in hospitals.

❖ The World Health Organization recently identified illegal and clandestine abortion as the leading cause of pregnancy-related deaths of young women in Latin America. Post abortion complications also now take up as many as 45% of the hospital beds in maternity wards in many areas. The average fertility rates range from 24 to 38 births per 1000 population in the Latin American region, as compared to from 13 to 16 births per 1000 in Western Europe and the United States. *"Advocates of women's rights in this region cite the woman's lowly place in society for official indifference toward maternal illness and death. Government officials argue that the Roman Catholic Church has prevented a change in the laws to provide safe abortions..."* (*WIN News*, Winter 1989)

❖ Natural Family Planning (NFP) techniques have come under criticism following a study that showed as many as 1 out of 3 women using such techniques may become pregnant in a given year. NFP techniques, such as periodic abstinence, the rhythm method, the basal body temperature method, the cervical mucus method, and the sympto-thermal method, are the only forms of family planning that are permitted by the Roman Catholic Church. Fortunately, according to a 1987 CBS News/New York Times poll, fully 89% of American Catholics believe it is possible to use "artificial" forms of birth control and still be good Catholics. (*Conscience*, Sept./Oct. 1988)

❖ And in the USA: *"An estimated 1.8 million women are beaten each year in their homes. In 1985, there were an estimated 795,000 abused children between the ages of 3 and 17 living in two-parent households. According to these studies, men are the main perpetrators of domestic violence and commit 95% of all assaults on spouses. In 70% of households in which women are abused, the men also commit child abuse..."* (*WIN News*, Winter 1989; 187 Grant St., Lexington, MA 02173)

As defined by Reich, emotional plague characters are highly energetic but neurotic people who, instead of privately working out their problems, set themselves up as the standard of normality or as the moral or scientific authority. The plague character then tries to force everyone else to conform to their inadequacy. Emotional plague behavior is pointedly and actively destructive to the health and well being of others. Whereas the average neurotic person may be indifferent or prefer to "live and let live", an emotional plague character actively organizes others to persecute various expressions of natural life. Reich said that we all have emotional plague impulses, but we must also take the responsibility to recognize these impulses and limit their destructiveness. This section will focus on recent occurrences of the emotional plague and the people behind them.

❖ It has come to our attention that the French scientist Jacques Benveniste, recently attacked for his findings of an energetic principle at work in homeopathic dilution experiments, has been threatened with loss of his job. Earlier this July, Benveniste was ordered to "cease and desist" from doing further research on the controversial topic by the director of the French INSERM laboratory (Scientific Council of the National Institute for Health and Medical Research), where he works. This order followed on the heels of a shamelessly distorted attack upon his research by the editors of the British journal *Nature*, in which a representative of the American CSICOP organization participated. Benveniste was additionally "ordered" to cease making public statements about his research findings, which were recently confirmed by other scientists in France, the USA, and the Soviet Union. *(Science, 21 July 1989)*

❖ "Jews are not good Americans. They have no understanding of what America is. The Jews have been working against our national interest – I mean trespassers against our race – I think they should be punished. The above shocking quote is attributed to David Duke, newly enacted member of the Louisiana State Legislature, by the Klanwatch Project of Montgomery Alabama. Klanwatch is a project of the Southern Poverty Law Center, which has kept an eye on some of the most hateful and dangerous of America's neo-Nazi, Ku Klux Klan, and skinhead groups. It is directed by Morris Dees, who recently sued the United Klans of America on behalf of a black woman whose son was lynched by their members. A jury of southerners awarded this woman $7,000,000 in damages, effectively putting that Klan group out of action. We take Mr. Dees very seriously in his call of alarm on these questions. A recent report from Klanwatch contains the following shocking statistics: - 1988 reached a five year high in serious antisemitic crimes. - In 1988, skinheads operated in 24 states, and were linked to five murders and dozens of violent assaults against minorities and synagogues. - Books with the following titles are openly sold by a group called the National Association for the Advancement of White People: *The Holy Book of Adolf Hitler,* by Mattersby; *The Hoax of the Twentieth Century,* by Butz ("proving" that the holocaust was a fraud); *The Veale File,* by Veale ("exposing the one-sidedness of the Nuremberg War Crimes Trials ....); *Jews Must Live,* by Roth ("A Jew reveals shady business practices of Jews"; *Facts Are Facts* ("Excellent probe of the Jewish question ... "; *The Turner Diaries,* by Pierce *("We are in a war to the death with the Jew, who now feels himself so close to his final victory that he can safely drop his mask and treat his enemies as the 'cattle' his religion tells him they are ... No matter how long it takes us and no matter to what length ... we must go, we'll demand a final settlement. We'll go to the uttermost ends of the earth to hunt down the last of Satan's spawn [Jews].")*

❖ The editors of *Pulse* were recently confronted by some very depressing examples of hidden fascistic, antisemitic aspects within the "New Age" movement. Copies of "The Hoax of the Twentieth Century" were sent to our offices by one individual who had previously expressed an interest in orgonomy, "free energy", and cloudbusting. Another individual known to us for his work on ·wholistic living" invited Laboratory workers to a "Tax Seminar" lecture on "Alternative Economic Theories", as presented by a woman who also arranged "new age" tours into the ancient cities of the Andes, such as Machu Picchu. This speaker was well-coached, presenting "facts" about "Jewish bankers· being behind all of the world's wars and political assassinations; we were also informed that the Jews had perpetrated the holocaust upon themselves, to "better cover their tracks". When angrily confronted by us, she admitted not being able to document her "facts", but casually deferred to her "teachers": ("Hey, I'm only presenting an alternative way of looking at history") The central underlying theme of these dangerous stupidities clearly revealed a mystical face, of a fundamentalist Christian orientation, with the assertion that the "Jew is in league with the devil", "Satan's spawn". This drivel could be dismissed and quickly forgotten, were it not for the organized and growing nature of such hate groups. For example, the organization that had coached the "lecturer" presenting the "Tax Seminar/ Alternative Economic Theory" also supplied a lengthy bibliography and other materials. The hook used to draw people in was well-baited: "Save money on your taxes, and learn why you are in economic trouble .. A major disturbing fact about that lecture was that *only a very few of the well-educated people attending it, namely those associated with this Laboratory, were upset about what was being said!* As an example of the dangerousness of this kind of "blame the Jew" philosophy, we cite the recent case in Los Angeles, where skinheads had mercilessly beaten an Arab couple with their newborn infant, mistaking them for Jews because of their darker complexion. One of the skinheads was getting his baseball bat out of his car to "finish them "when the police arrived. In our letter of protest to the host of the meeting, we pointed out that educated persons who attended such a lecture were not likely go out and "bash Jews" afterwards. But a similar lecture given to young unemployed, uneducated skinheads, who are looking to blame someone for their own desperate and miserable situations, is likely to provoke just such a reaction. The silence of the well-educated is an implicit approval of such violence. As Reich pointed out in the *Mass Psychology of Fascism* (an analysis of the sex-economic origins of fascism in Germany during the 30's), mysticism and intellectualism often mask a deeper, violent rage.

❖ More sickening than the above, is the fact that we have recently heard of people who are promulgating the above kind of racist hatred under the guise of orgonomy, and in the name of Reich. For example, we have recently been informed of the existence of neo-Nazi groups in Germany, who want to develop a "National Orgonomic Front". Likewise, in America, a man who regularly shows up at various "psychotronic" and "new age" festivals as a representative of the orgone energy blanket, is a self-proclaimed white supremacist.

❖ According to Dr. Gary Null, who hosts the *Natural Living* radio talk show in New York City, *"almost 200 doctors practicing preventative medicine as a healing modality are being attacked"* by private medical or governmental organizations. Many are forced to relocate outside of the USA, where they have more freedom. *"In exposing the 'greed and ignorance' surrounding orthodox medicine, Null goes so far as to call the medical establishment 'the biggest cult in America.' He also accuses the media of news blackouts, citing many pertinent discoveries that never get published in the medical journals. Null asks journalists to question and challenge facts, find out where the money for research comes from, and to stop being the tools of manipulation. "(Organica,* Summer 1989)

❖ Regarding the organization CSICOP (Committee for the Scientific Investigation of Claims of the Paranormal), we recently received a paper titled "CSICOP and Skepticism: An Emerging Social Movement", by George P. Hansen, which revealed the following facts about CSICOP:

- The total paid circulation of its journal *Skeptical Inquirer,* is at least 25,000, and the group is emerging as a considerable social force.

- When formed in 1978, it was primarily a scholarly body, but soon took a denunciatory approach.

- A major priority of CSICOP is influencing public opinion, through use of various media contacts. Pressure has also been brought to bear against newspapers and TV networks that give favorable treatment to unusual topics, and also against universities and schools that present paranormal subject material in a balanced or sympathetic manner.

- Early on, CSICOP undertook an investigation of astrological claims, finding that there was some evidence and statistical validity to some of those claims. However, CSICOP continued to attack astrologers and those scientists honestly researching the subject. They furthermore refused to publish their own findings, leading to a bitter schism, with charges of "cover-up", and the resignation of many scientist members.

- CSICOP presently has an established policy of not doing research.

- The group does not, almost as a policy, allow persons attacked in its pages the right of rebuttal or response to criticism, and it has been sued on numerous occasions.

- Many local "skeptics" societies have been formed in the USA, with CSICOP acting as the central coordinator. These groups "watchdog" their local areas, and engage in letter-writing actions and media campaigns against unorthodox scientists and paranormal researchers, and have often themselves been sued.

- Its membership is represented by a disproportionate percentage of magicians, though a few well-known scientists are members, or lend their names to the group, and are prominently listed on its promotional materials (such as Carl Sagan, Stephen Jay Gould, B. F. Skinner, F.H.C. Crick, L. Sprague de Camp, Paul MacCready, James E. Oberg, and Isaac Asimov).

- Many of its members hold strong religious views which are antagonistic to the paranormal, and this provides the primary motivation for many of the positions the group takes, and for its published articles. (Contact: George Hansen, Psychophysical Research Laboratory, 301 College Road, East, Princeton, NJ 08540.)

❖ The Toronto *Globe and Mail* newspaper discontinued David Suzuki's science column most recently, saying that it took too much of an "environmental focus". This is what Suzuki said: *"If we environmentalists care about saving anything, we have to throw our lot in totally with native land claims. If we succeed, we will save the last vestiges of our wilderness and, in the process, we will come to realize that there is a radically different way of looking at the world. "*

❖ *'Official science [is] a malignant form of science whose purpose is to support official policies instead of objectively assessing their scientific truth or falsity.* "- Dr. Irwin J. Bross, who once directed a Department of Energy study into the health hazards of low level radiation; Bross demonstrated that low level radiation had a deadly effect upon human life, and was subsequently fired for his honesty in reporting this fact. *(Conscious Living,* May 89)

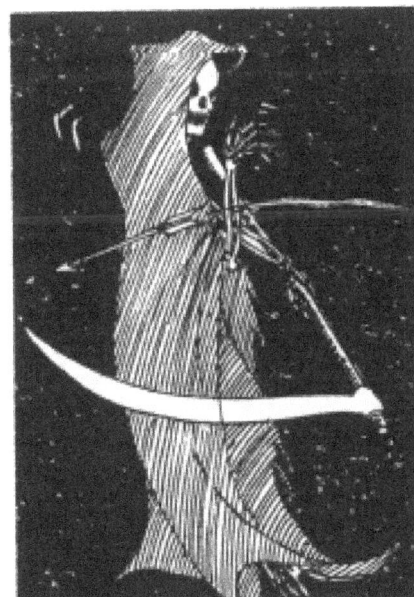

CLIMATE FEATURES AND UNUSUAL PHENOMENA

Editor's Comments: *There are several interesting patterns which were observed in global climatic phenomena during the six months of study reported here:*

1. There is a tendency for episodes of drought to occur in diffferent places around the world at the same time, with a subsequent termination of drought conditions in a generally simultaneously manner.

2. The large solar flare of early March was followed by an energizing of the Earth's weather systems, such that tendencies towards stagnation were dramatically reduced. The pattern appears to be: "Storms on the Sun, Storms on the Earth".

3. A heat-wave episode developed in the Western USA in early July following several late June underground nuclear bomb tests in Nevada. The agitated atmosphere, identified as "oranur" by Reich, slowly moved eastward across the USA, eventually to appear in Europe. Europe had previously experienced heat wave and drought conditions, but these tendencies appear to have been enhanced by the appearance of the excited atmosphere.

4. The most interesting phenomena of the period was the development of super wet conditions in the arid desert interior portions of Australia, mostly during March. In fact, many parts of the Australian Desert received over 1000% of their normal rainfall for that month. The eastern portions of Australia continued to receive additional record rains in subsequent months, but the western portions were thereafter gripped by an intense drought. The following is a news report which appeared in the San Francisco Chronicle *on May 27, 1989:*

'Big Wet' Turns Outback Into 'Garden of Eden'
by Robert Woodward, Reuters, Wilpena Pound, Australia.

The Australian outback is usually a semiarid scrubland stretching to the shores of a dry salt lake. But one of the wettest spells since Europeans reached the area in the mid-19th century has completely transformed the desert heart of the world's driest continent. The outback is now miles of golf course-like green merging into Lake Torrens, which is full for the first time in recorded history.

"The Big Wet has turned a dust bowl into the Garden of Eden", says Brent Williams, wildlife ranger in Flinders National Park. Parched outback areas in Queensland, South Australia and the Northern Territory were deluged with a foot of rain in March, three times the annual average for the Red Center. Creeks, dry for yrears on end, turned overnight into torrents hundreds of yards wide, flooding huge areas. Much of the floodwater headed for the sprinkling of salt lakes in the north of South Australia and in particular Lake Eyre, which attracts water from an area roughly the size of France, Spain and Portugal. Two months after the downpour, Lake Eyre, normally covered by a thick salt crust and lying 50 feet below sea level, has become the world's 16th biggest lake after filling for only the fourth time since white settlers first discovered it.

Pelicans, terns and other birds have flocked in from the coast in the thousands to breed on the lakes, which should stay full for several months, and perhaps a year or more, after further heavy rain in sothern Queensland over the past month. The rain reaps huge, sudden benefits. Last month, rangers found more than 100,000 nests of the banded stilt, a wading bird thought to be extinct in South Australia, on an island in flooded Lake Torrens. Food for the birdlife is provided by freshwater fish--bream, gudgeon and perch--washed down flooded rivers to the lakes, which become breeding grounds for every kind of insect. The water also causes extraordinary happenings further down the food chain. The waterholding frog is awakened after spending many years cocooned in an underground cell, and shrimps, crabs and plankton appear--from where remains a mystery. Bushes, grasses and trees are showing luxuriant growth while flowers, normally dormant as winter approaches, are growing at breakneck speed. Farmers, hit by several years of drought conditions, have also profited from the downpour... However, the rain has also brought its share of problems... floods washed away much precious topsoil and... Darling Downs, the nations breadbasket, is suffering from a plague of mice.

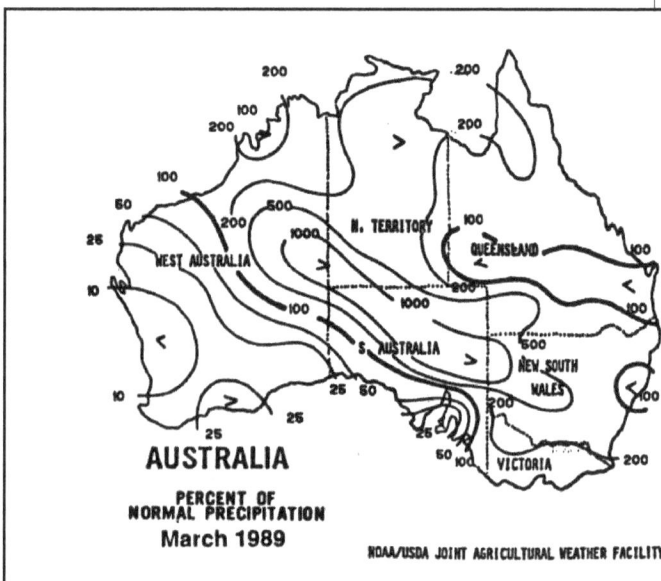

AUSTRALIA
PERCENT OF NORMAL PRECIPITATION
March 1989
NOAA/USDA JOINT AGRICULTURAL WEATHER FACILITY

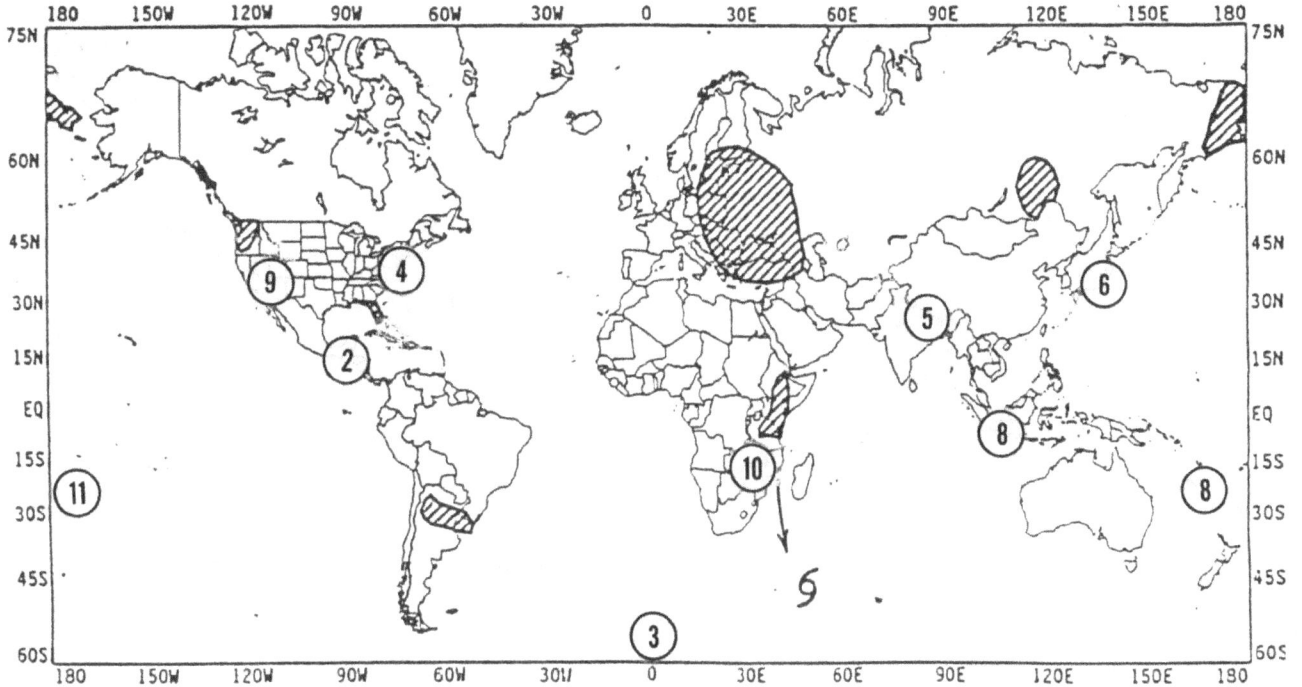

FOR THE WEEK ENDING MARCH 11, 1989
(Continued from Vol. 1, No. 1 of *Pulse of the Planet*)

Persistent Conditions (shaded)

NW USA: Heavy rains bring relief; long term drought deficit remains (9 weeks).

FLORIDA: Rains end dry spell. (Ended at 8 weeks.)

URUGUAY & N. ARGENTINA: Heavy rains reported; area still warm and dry (15 weeks).

E. EUROPE & MIDDLE EAST: Dryness ends (ended at 12 weeks); warmth persists (9 weeks).

E. AFRICA: Dryness spreads (4 weeks).

SIBERIA: Above normal temperatures prevail in south central and eastern regions (22 weeks).

ITALY: Drought; 30% of crops in Lombardy may be destroyed.

Transient Events (numbered)

(1) SOLAR FLARE: Disrupts shortwave and navigational communications.

(2) VOLCANO: GUATEMALA; plantations destroyed, hundreds evacuated.

(3) ICEBERGS: Two Antarctic islands, Terra Novas, listed on the maps since 1961, are actually huge icebergs.

(4) WINTER STORM: Eastern seaboard battered; 14 in. of snow fell in Missouri in one day.

(5) ELEPHANT RAMPAGE: NEPAL; Seven wild elephants drank home-brewed liquor; destroyed homes,

bridges, and killed two people.

(6) Mar. 6: EARTHQUAKE; Japan, 5.9 mag.

(7) Mar. 7: NEW MOON; Partial eclipse of Sun.

(8) Mar. 8: EARTHQUAKE; Vanuatu Isl., 6.2 mag.; Indonesia, 6.0 mag.

(9) Mar. 9: ATOMIC BOMB; U.S.A., Nevada, 20-150 KT.

(10) Mar. 10: EARTHQUAKE; Malawi, 6.2 mag., nine people killed, 100 injured, 50,000 left homeless.

(11) Mar. 11: EARTHQUAKE; Tonga Isls., 6.3 mag.

Solar Geomagnetic Data

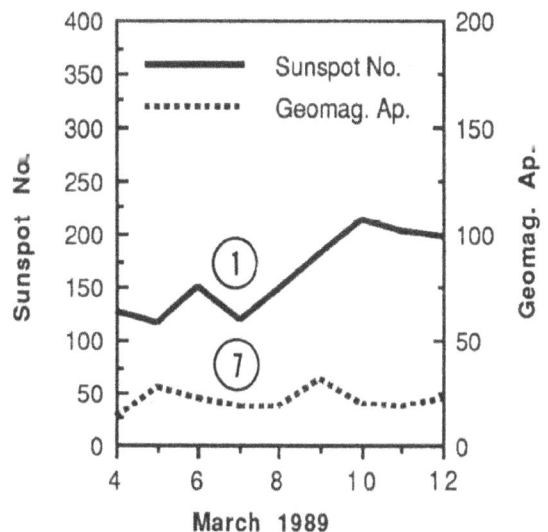

March 1989

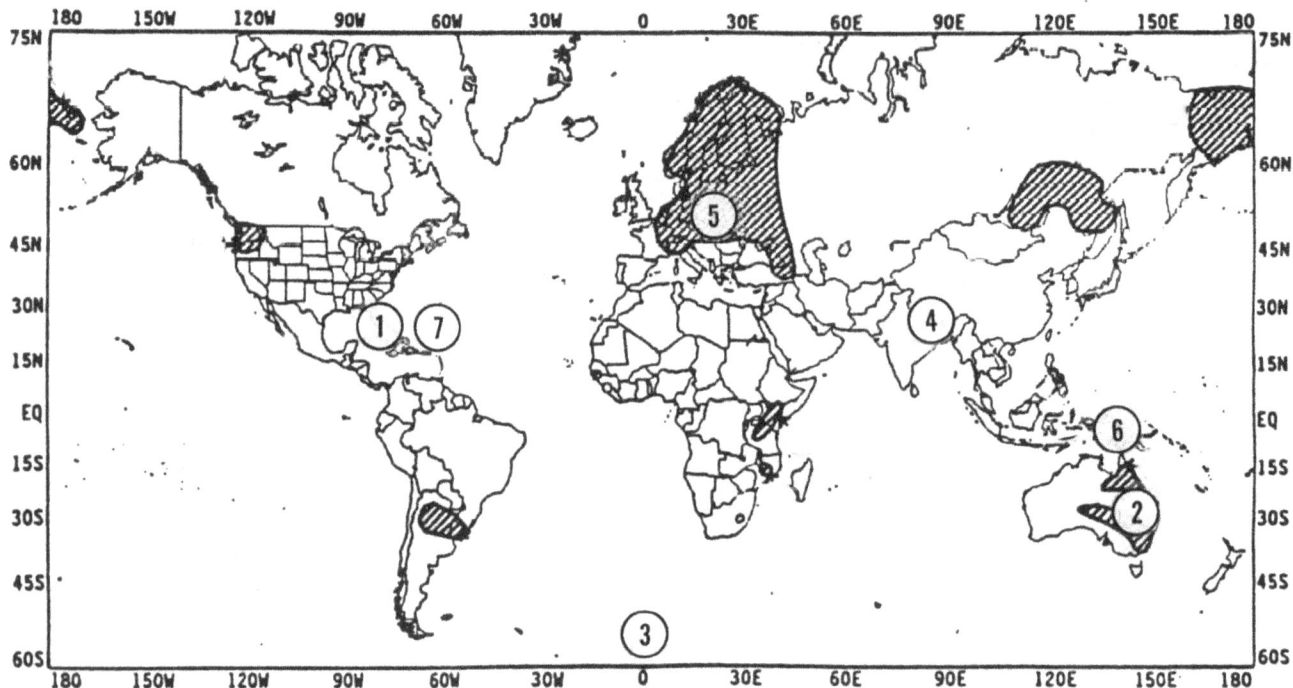

FOR THE WEEK ENDING MARCH 18, 1989

Persistant Conditions (Shaded)

NW USA: Heavy rains continue to bring relief from dryness (ends at 9 weeks).

URUGUAY, N. ARGENTINA: Heavy rains reported; warm
spell ends (at 15 weeks); most areas still experiencing long term dryness (38 weeks).

E. EUROPE, MIDDLE EAST: Mild, above normal temperatures persist (10 weeks).

E. AFRICA: Unusually dry conditions continue (5 weeks).

SIBERIA: Mild, above normal temperatures persist in S.
Central and East.

AUSTRALIA: Torrential rains; 11.8 inches in some areas.

Transient Events (numbered)

(1) MERCURY CONTAMINATION: FLORIDA; Game fish
in Everglades swamp have four times the safe amount
of methyl mercury; worst ever in USA; source may be
illegal waste dumping from electronics plants.

(2) FLOODS: AUSTRALIA; vast areas of outback turned
into inland seas by record rains; hundreds stranded;
wheat farmers welcome rains.   MALAWI: Floods
devastate South, 50,000 homeless.

(3) ICEBERG: ANTARCTICA; iceberg size of Hong Kong,
largest ever; broke away from Ross ice shelf 18
months ago; traveling in 620 mile circle.

(4) MOSQUITO INVASION: INDIA; airline cockpit in-
vaded by swarm in Calcutta; flight delayed.

(5) EARTHQUAKES: E. GERMANY; 5.5 mag., set off by
underground mine blasting along border of E. and W.
Germany; buildings damaged.

(6) EARTHQUAKE: PAPUA NEW GUINEA, 5.7 mag.

(7) Mar. 13: SOLAR FLARE; Massive solar flare of last
week causes northern lights as far south as Jamaica;
increased upper atmospheric temp. by 2,000 deg. F.

Solar Geomagnetic Data

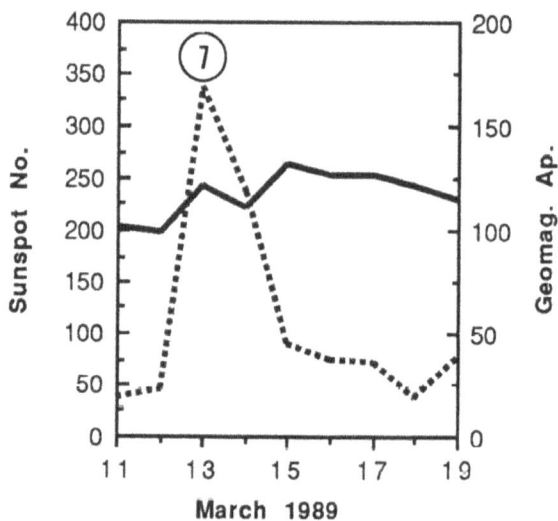

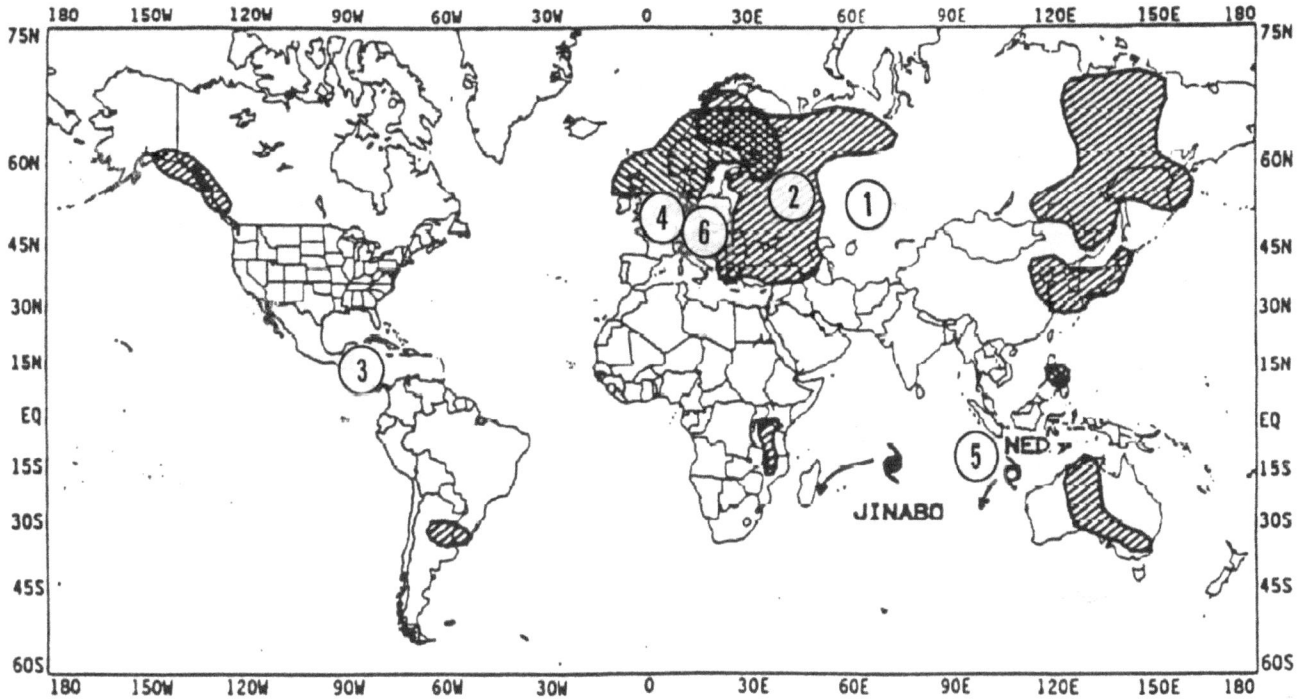

FOR THE WEEK ENDING MARCH 25, 1989

Persistent Conditions (shaded)

BRITISH COLUMBIA, ALASKA: Dry conditions develop in coastal areas (5 weeks).

URUGUAY, N. ARGENTINA: Heavy rains end short-term dryness (ended at 39 weeks).

SCANDINAVIA, NW EUROPE: Above normal precipitation (4 weeks).

E. EUROPE, MIDDLE EAST: Abnormally mild weather continues (11 weeks).

MALAWI, TANZANIA, KENYA: Rains provide relief for dry areas; Malawi flooded (ended at 6 weeks).

SIBERIA: Unusually mild conditions continue in cental and eastern areas (24 weeks).

E. CHINA, N. and S. KOREA, JAPAN: Region turns abnormally warm and wet (5 weeks).

CENTRAL PHILIPPINES: Above normal rains continue to fall during dry season (4 weeks).

AUSTRALIA: Heavy rains continue to fall; vast areas of desert sprout new growth (2 weeks).

MIDWEST USA: Droughts; 75 % of Kansas winter wheat crop is lost; farmers try to save soil; also in OKINAWA, NEW YORK CITY.

Transient Events (numbered)

(1) GLACIER WARNING: USSR; Large glacier moving at 3 yards per day, threatens populated valley in Pamirs.

(2) RADIOACTIVE: USSR; A quarter million Soviets on land contaminated by Chernobyl; food is imported.

(3) WILDFIRES: NICARAGUA; 75,000 acres of forest burning out of control.

(4) TOXIC PESTICIDE: ENGLISH CHANNEL; Five-ton container of Lindane pesticide sunk in Channel; marine life wiped out if it leaks.

(5) TROPICAL STORMS: *Ned* curves harmlessly around Australian coast; another forms near Madagascar.

(6) EARTHQUAKES: GREECE; 5.8 mag.

(7) Mar. 20: SPRING EQUINOX; Equal hours of day and night.

(8) Mar. 22: FULL MOON.

Solar Geomagnetic Data

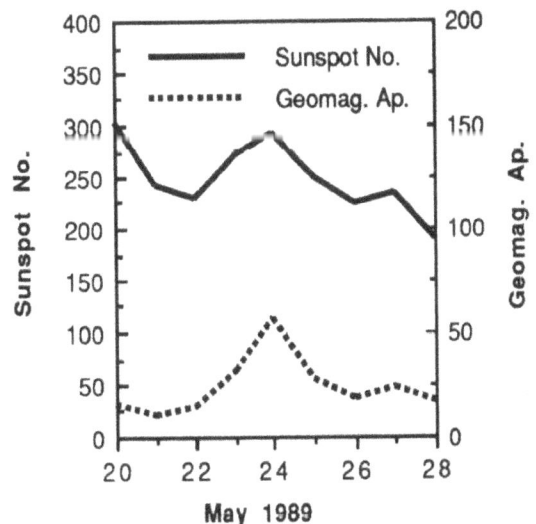

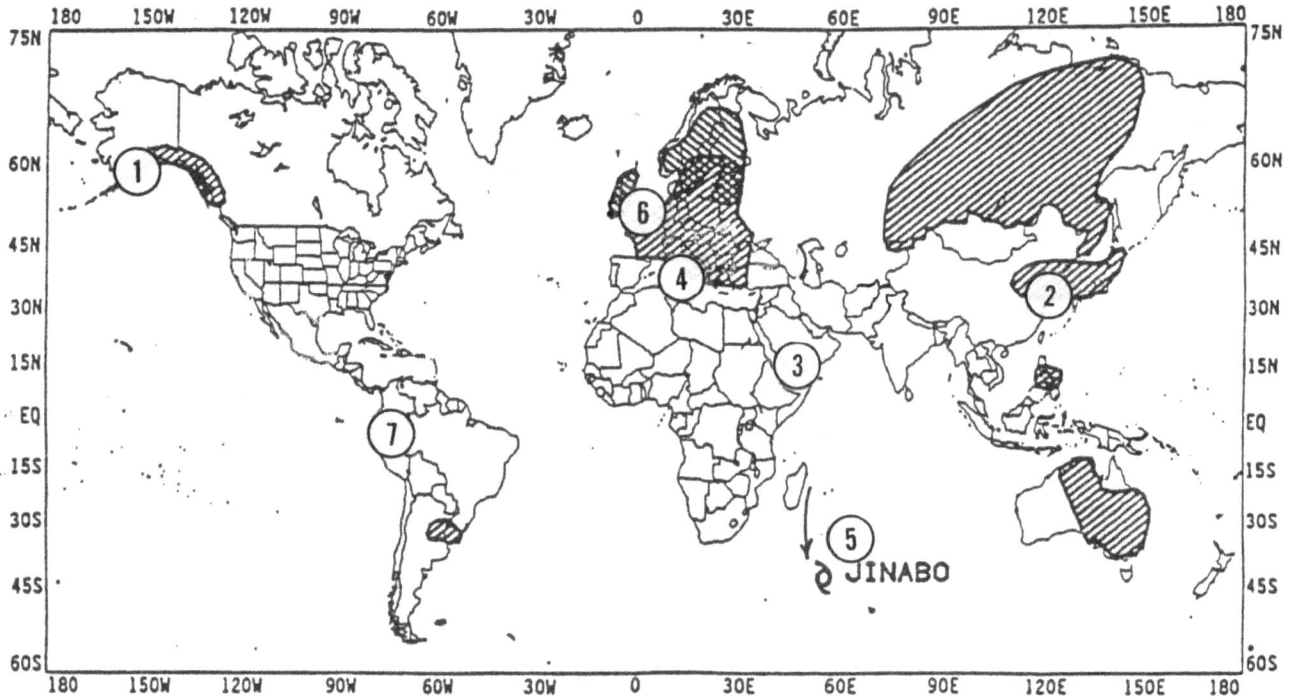

FOR THE WEEK ENDING APRIL 1, 1989

Persistant Conditions (shaded)

BRITISH COLUMBIA, ALASKA: Dry conditions persist in coastal areas (6 weeks).

URUGUAY, E. ARGENTINA: Rains end; dry conditions return (40 weeks).

N. EUROPE: Heavy rains continue (5 weeks).

S. CENTRAL EUROPE: Abnormally mild conditions persist (12 weeks).

SIBERIA: Unseasonably mild conditions continue (25 weeks).

E. CHINA, KOREA, JAPAN: Unusually high temperatures prevail (7 weeks); rains ease up (ends at 6 weeks).

PHILIPPINES: Above normal rains continue (5 weeks).

AUSTRALIA: Large quantities of rain continue to fall (3 weeks).

BRAZIL, SRI LANKA: Droughts.

Transient Events (numbered)

(1) OIL SPILL: ALASKA; 240,000 barrels of crude oil spill into Prince William Sound from Exxon oil tanker that ran aground. Beaches blackened, wildlife dying, salmon run and hatcheries endangered.

(2) OCEAN HEATING: CHINA; Sea level and ocean surface temperatures rising in coastal areas; storm flooding threatens populations.

(3) FLOODS: SOUTH YEMEN, SOMALIA; Torrential rains cause damage; 34 people killed.

(4) VOLCANO: SICILY; Two volcanic blasts shook island of Stromboli.

(5) TROPICAL STORMS: *Jinabo* heads south from Madagascar.

(6) "PSYCHOTIC" BEARS: ENGLAND; Two polar bears psychotic from zoo imprisonment and boredom; they receive "toy and food psychotherapy", but no freedom.

(7) EARTHQUAKES: PERU, 5.6 mag.

(8) APRIL 1: ALL FOOL'S DAY: LOCH NESS, SCOTLAND: Lake monster stages hunger strike for all imprisoned zoo animals; vows to "disappear forever" unless humans behave more like animals.

Solar Geomagnetic Data

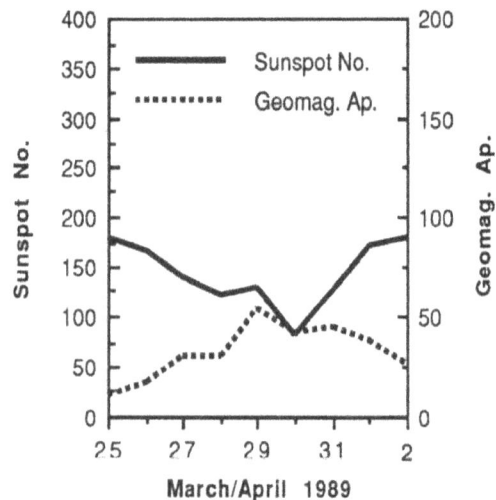

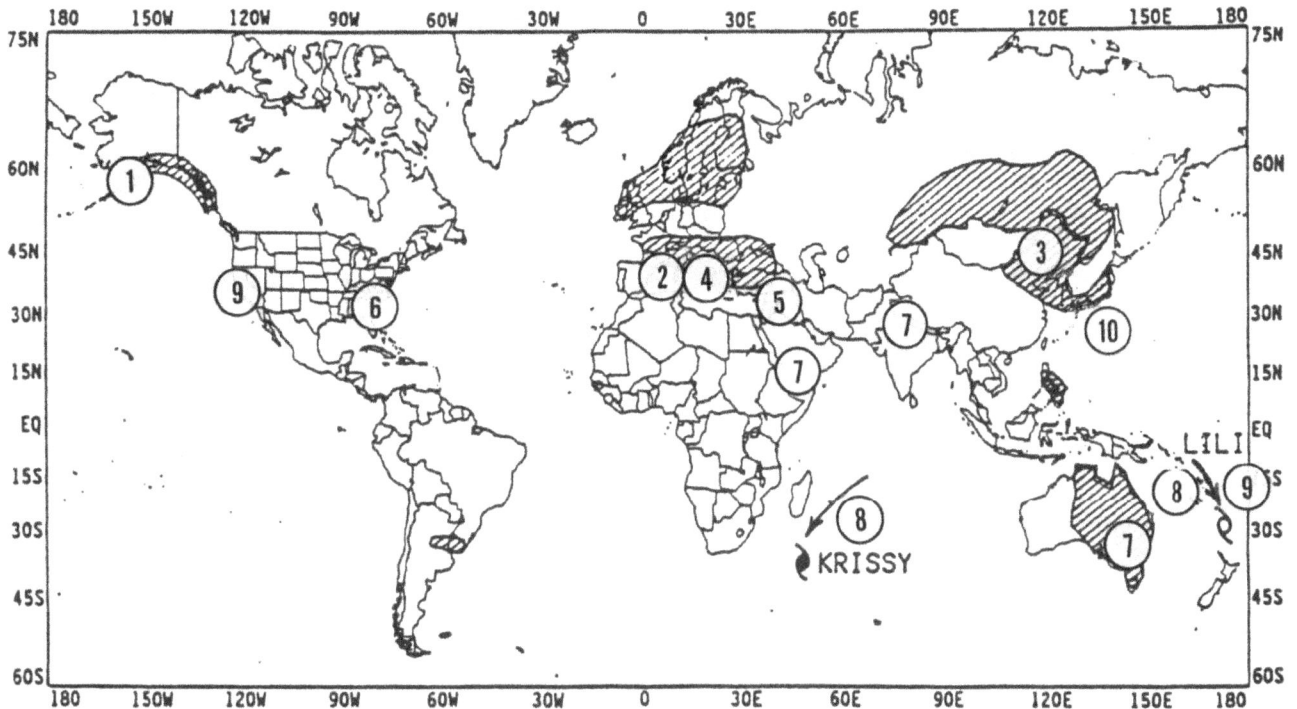

FOR THE WEEK ENDING APRIL 8, 1989

Persistent Conditions (shaded)

BRITISH COLUMBIA, ALASKA: Very dry weather continues in coastal areas (7 weeks).

URUGUAY, E.ARGENTINA: Area still dry (41 weeks)

N. EUROPE: Wetness diminishes (ends at 5 weeks).

S. CENTRAL EUROPE: Unusually warm conditions persist (13 weeks).

SIBERIA: Abnormally mild weather continues (26 weeks).

E. CHINA, KOREA, JAPAN: Above normal temperatures prevail (8 weeks).

PHILIPPINES: Wetness eases (ends at 6 weeks).

AUSTRALIA: More heavy rains fall (4 weeks).

Transient Events (numbered)

(1) OIL SPILL SPREADS: ALASKA; Raw, crude oil spreads into Gulf of Alaska; toll to wildlife mounts.

(2) WILDFIRES: S. FRANCE, CORSICA; Caused by months of drought.

(3) WILDFIRES: CHINA; 40 sq. miles of grasslands blaze.

(4) SMOG: GREECE; Athens registered highest levels ever of air pollution.

(5) TOXIC CLOUD: LEBANON; Beirut bombing of oil terminal creates 8 sq. mile toxic cloud, traveling north; if cloud fell before rains dissolved it, entire forests endangered, people's eyes and lungs could be burned.

(6) TORNADOES: ALABAMA to FLORIDA; Two days of severe weather in region.

(7) FLOODS: AUSTRALIA, YEMEN, INDIA.

(8) TROPICAL STORMS: Typhoon hits Australia; other storms form near Madagascar and the Soloman Islands.

(9) Apr. 2: EARTHQUAKE; VANUATU ISLS., 6.3 mag.

(10) Apr. 3: EARTHQUAKE; CHICHI JIMA, 5.2 mag.

(11) Apr. 6: NEW MOON.

Solar Geomagnetic Data

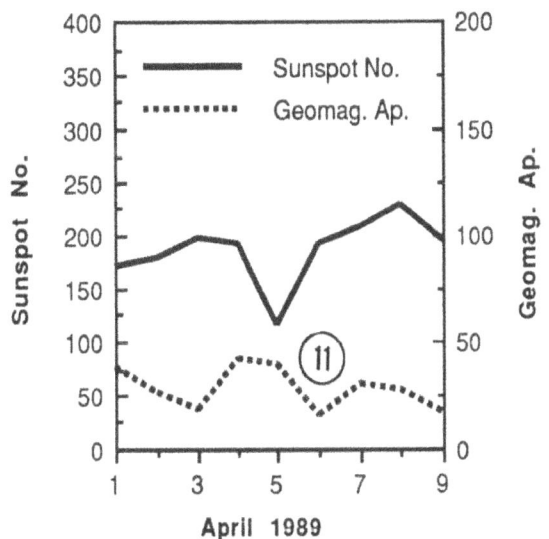

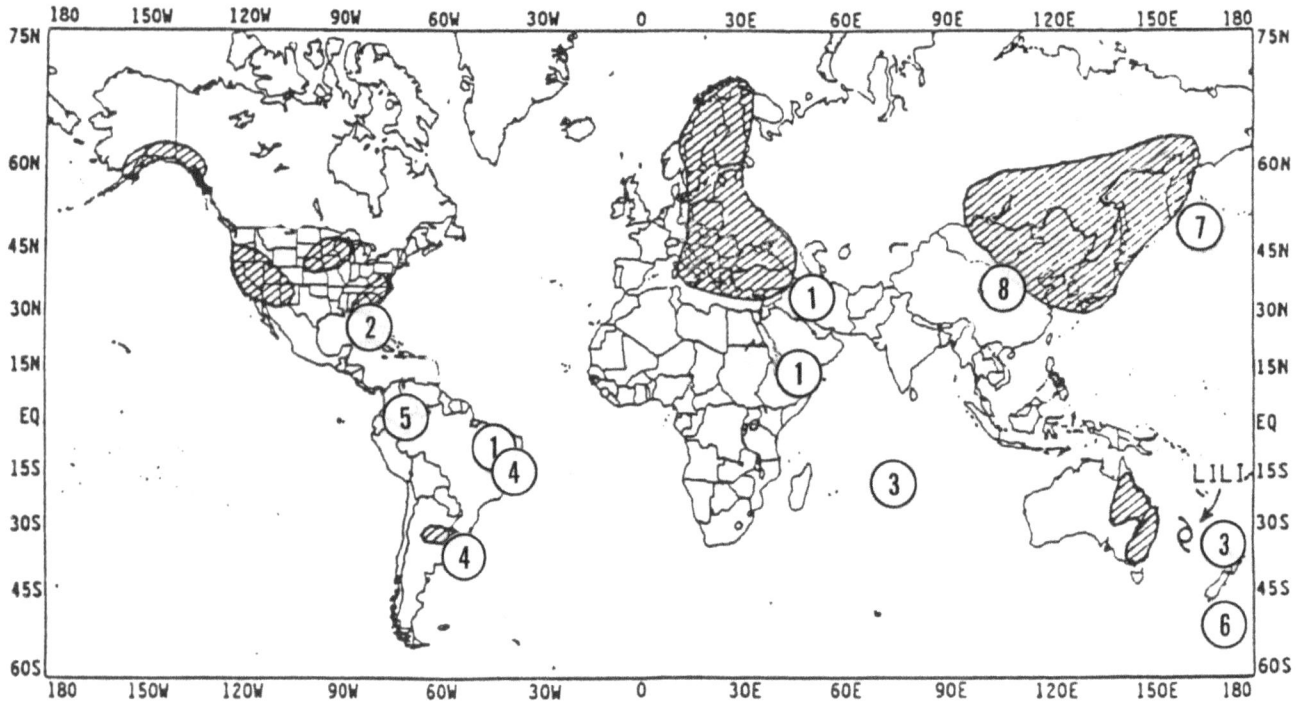

FOR THE WEEK ENDING APRIL 15, 1989

Persistent Conditions (shaded)
BRITISH COLUMBIA, ALASKA: Dry weather continues in coastal sections (8 weeks).
W. USA: Unseasonably warm conditions persist (8 weeks).
CENTRAL USA: Very dry conditions develop (4 weeks).
SE USA: Unusually cold Canadian air invades (2 weeks).
URUGUAY, E. ARGENTINA: Dryness persists (42 weeks).
S. CENTRAL EUROPE: Above normal temperatures continue (14 weeks).
E. ASIA: Abnormally high temperatures continue (9 weeks).
AUSTRALIA: "Big Wet" continues (5 weeks).
DROUGHTS: BANGLADESH, SRI LANKA, THAILAND.

Transient Events (numbered)
(1) FLOODS: BRAZIL, DJBOUTI, IRAN
(2) WILDFIRES: FLORIDA; 18,000 acres burned in southern section.
(3) TROPICAL STORMS: One passes over New Caledonia; another forms in the Indian Ocean.
(4) HEAVY RAINS: BUENOS AIRES, BRAZIL & ARGENTINA.
(5) LANDSLIDE: COLUMBIA; Village damaged; early reports falsely claimed a meteor hit.
(6) WASP WARNING: NEW ZEALAND; Swarming wasps migrate north; will attack people.
(7) Apr. 11: EARTHQUAKE; USSR, Kuril Isls., 6.5 mag.
(8) Apr. 15: EARTHQUAKE; CHINA, Sichuan Province, 6.2 mag.; 11 killed, 42 injured, considerable damage.

Solar Geomagnetic Data

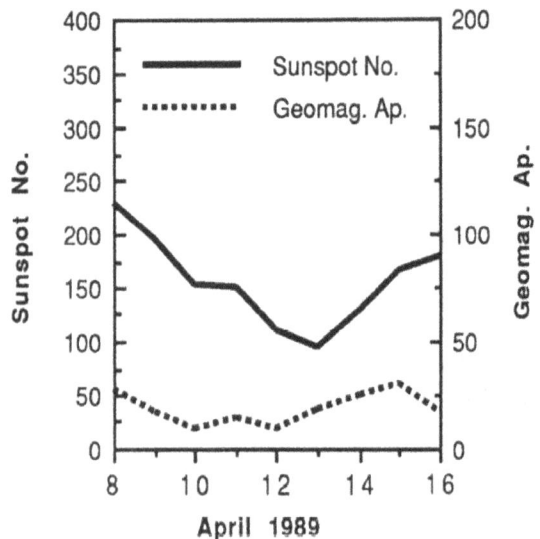

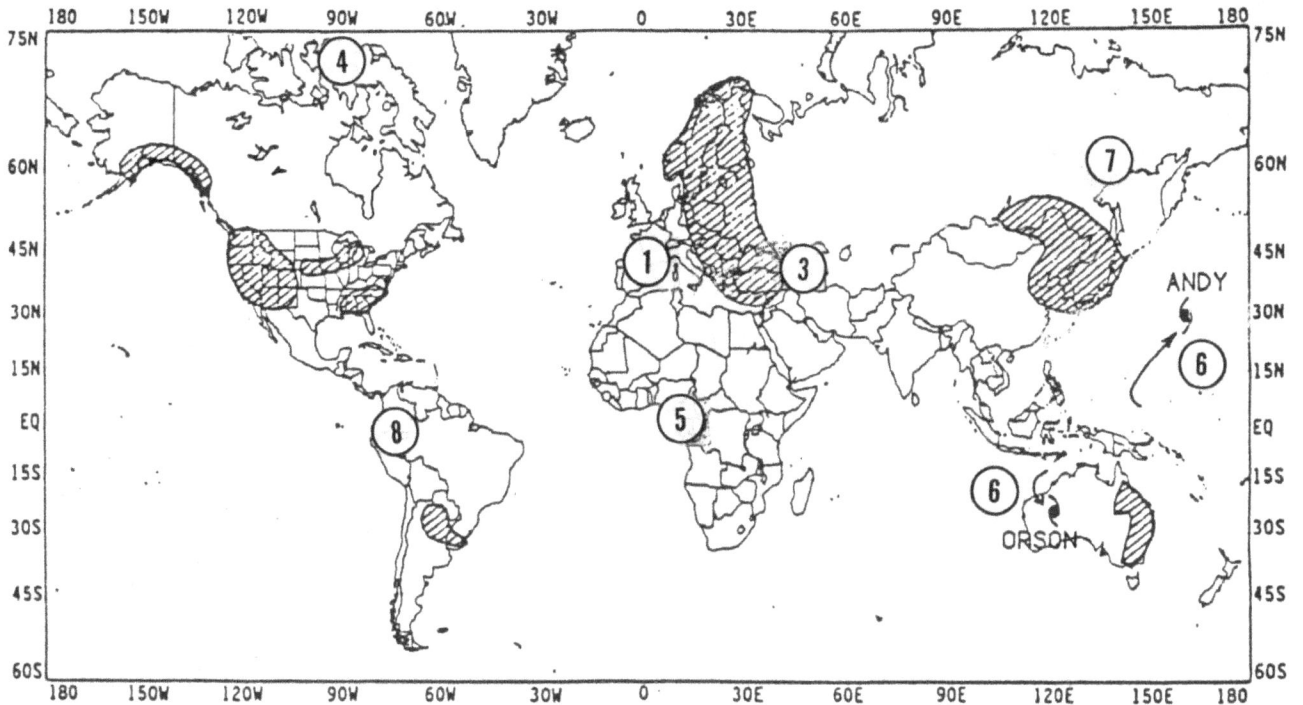

FOR THE WEEK ENDING APRIL 22, 1989

Persistent Conditions (shaded)

BRITISH COLUMBIA, ALASKA: Dry weather persists in coastal area (9 weeks).

W. USA: Above normal temperatures continue (9 weeks).

CENTRAL USA: Dryness persists; some rain in Iowa and Illinois (5 weeks)

SE USA: Seasonal temperatures return (end at 2 weeks).

URUGUAY, E. ARGENTINA: Some rain falls; moisture deficits continue (43 weeks).

EUROPE, MIDDLE EAST: Area still warm (15 weeks).

E. ASIA: Unusual warm weather ends in Siberia (ended at 27 weeks); Japan, Korea, and NE China are still above normal (10 weeks).

AUSTRALIA: "Big Wet" persists; heavy rains continue (6 weeks).

OCEAN WARMING: Temperature of world's oceans rose on average 1/5 of a degree F since 1982. NOAA.

Transient Events (numbered)

(1) HEAVY RAIN: S. EUROPE; Storms from Spain to Portugal eased drought and forest fire danger.

(2) SOLAR CONNECTION: Soviets announce that eruptions of mud volcanoes may be linked to solar storms; both occurred simultaneously.

(3) FLOODS: SOVIET GEORGIA; Landslide causes river to back up; dozens of people killed.

(4) STRANDED: ARCTIC CANADA; Solar storm blocked radio communications from a team measuring pollution in arctic; supplies eventually restored.

(5) THUNDERSTORM: S.GABON: Walls of prison collapse, but shocked prisoners do not escape.

(6) TROPICAL STORMS: *Orson* hits Australia; *Andy* heads toward Guam.

(7) Apr. 20: EARTHQUAKE; SOVIET ASIA; 6.5 mag.

(8) Apr. 21: FULL MOON.

Solar Geomagnetic Data

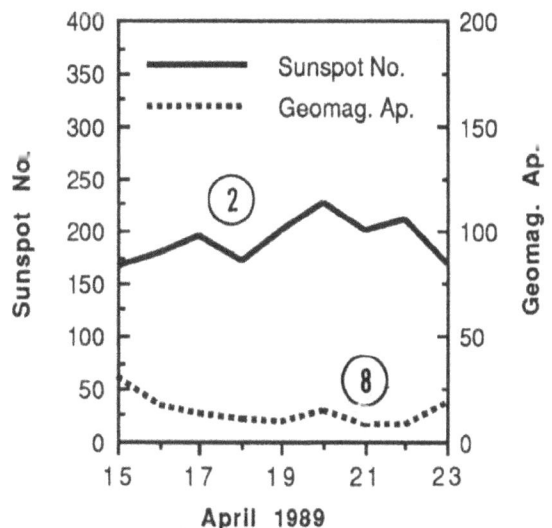

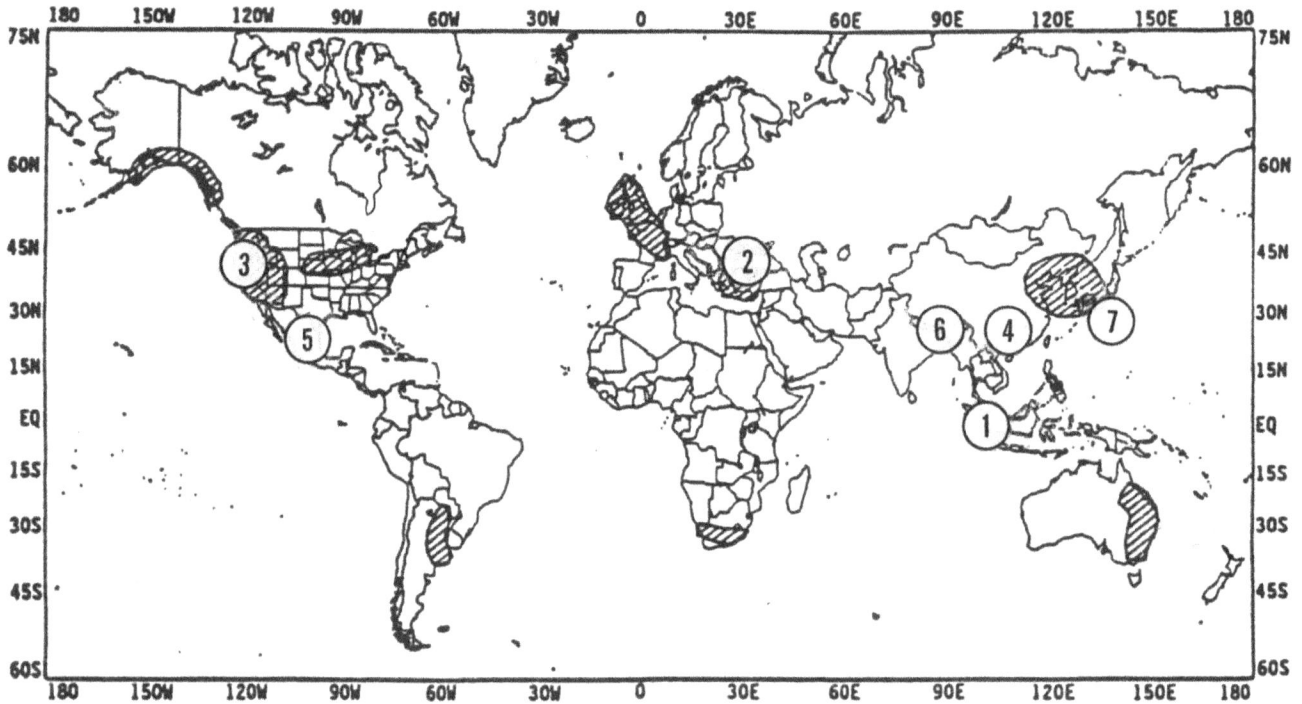

FOR WEEK ENDING APRIL 29, 1989

Persistent Conditions (shaded)

BRITISH COLUMBIA, ALASKA: Coastal region remains dry (10 weeks).

W. USA: Warm conditions end (ended at 9 weeks).

CENTRAL USA: Rains end dry spell (ended at 6 weeks).

ARGENTINA: Some rain eases dryness; moisture deficit continues (44 weeks).

EUROPE, MIDDLE EAST: Cold air ends warm conditions in Europe (ended at 16 weeks); Turkey and Middle East still warm.

S. AFRICA: Unusually heavy precipitation (4 weeks).

E. ASIA: Temperatures cool to normal (ended at 10 weeks).

E. AUSTRALIA: Abnormally wet weather persists (7 weeks).

LOCUSTS: AFRICA, MIDDLE EAST; Yearlong plagues die out.

Transient Events (numbered)

(1) VOLCANO: INDONESIA; Soputan on the island of Sulawesi erupts; no damage reported.

(2) NATURAL GAS: BLACK SEA; Soviet scientists trying to clean sea of combustible hydrogen sulfide and use it in power plants. The gas is creating a "dead zone" at a depth of 240 ft. and it is slowly rising.

(3) FALLING ICE: OREGON; Volleyball size chunk of ice fell through roof of Portland house; chunk may have formed on wing of airplane in a thunderstorm.

(4) EARTHQUAKES: CHINA, 6.7 mag.

(5) Apr. 25: EARTHQUAKE; MEXICO, Guerrero, 6.9 mag.; 3 people killed, several injured.

(6) Apr. 26: THUNDERSTORMS: BANGLADESH; Storms and tornadoes sweep into area after townspeople pray for an end to drought.

(7) Apr. 27: EARTHQUAKE; JAPAN, South of Honshu, 6.1 mag.

Solar Geomagnetic Data

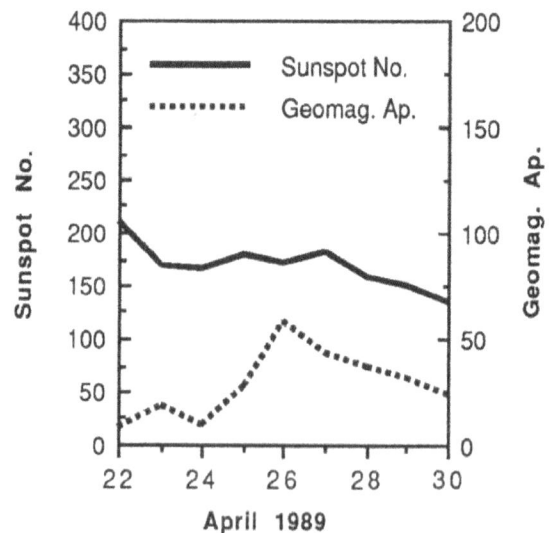

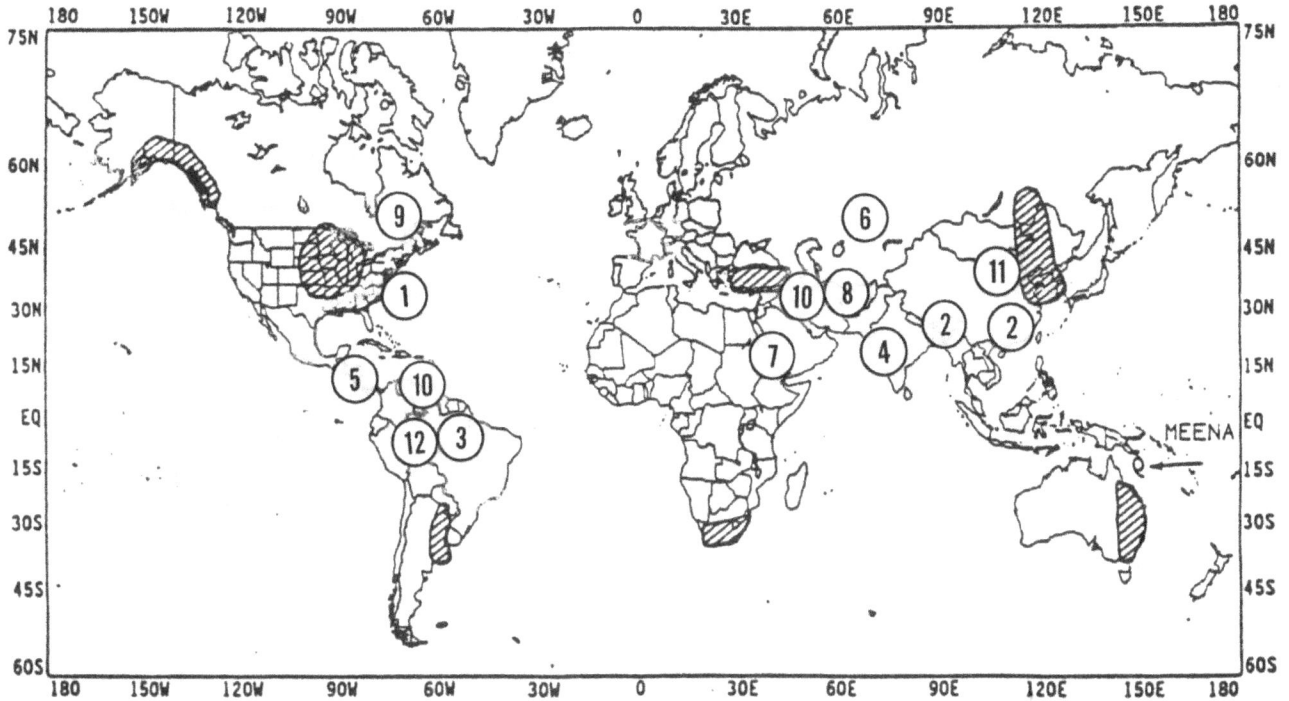

FOR THE WEEK ENDING MAY 6, 1989

Persistent Conditions (shaded)

BRITISH COLUMBIA, ALASKA: Coastal areas are still dry (11 weeks).

CENTRAL USA: Dry conditions return (7 weeks).

ARGENTINA: More rain, southern part still dry (45 weeks).

TURKEY: Hot, dry conditions prevail (8 weeks).

S. AFRICA: Wetness diminishes, isolated showers (5 weeks).

E. ASIA: Above normal temperatures return (11 weeks).

E. AUSTRALIA: Rain eases in some areas (8 weeks).

Transient Events (numbered)

(1) HEAVY RAINS: USA; The South and Atlantic coast inundated with rain.

(2) THUNDERSTORMS: BANGLADESH; New storm hits area devastated by tornadoes; CHINA; Shanghai and Hong Kong hit hard; 6000 injured.

(3) HEAVY RAINS: BRAZIL; Mudslide buried mining site in Amazon; 30 gold miners killed.

(4) TORNADO: INDIA; 10 killed, 100 injured in east Indian state of Orissa.

(5) VOLCANO: COSTA RICA; Poas volcano erupts, creating new crater.

(6) SPRING SNOWSTORM: SOVIET UZBEKISTAN; Heavy snow followed by hail, high winds, and heavy rains; livestock killed, crops and homes damaged.

(7) OIL SPILL: RED SEA; 15,000 barrels of crude oil dumped when Indian tanker hits reef; spill contained by Saudis' clean-up operation.

(8) BUTTERFLY INVASION: IRAN; Millions of brown butterflies swarmed in northeast town; dry spring may be responsible for migration.

(9) CORMORANT KILL: QUEBEC; Bird population has doubled, threatens other wildlife; 10,000 to be killed, 60-70% of eggs to be destroyed.

(10) EARTHQUAKES: VENEZUELA, 5.9 mag.; IRAN, 5.5 mag.

(11) May 3: EARTHQUAKE; CHINA, Sichuan Province, 6.1 mag.

(12) May 5: EARTHQUAKE; W. BRAZIL; 6.4 mag.

(13) May 5: NEW MOON.

Solar Geomagnetic Data

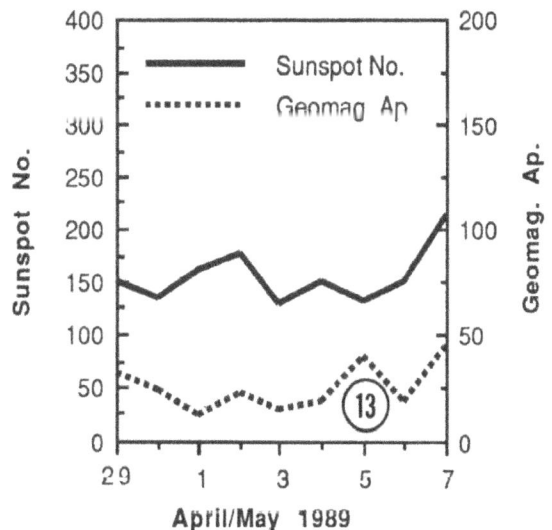

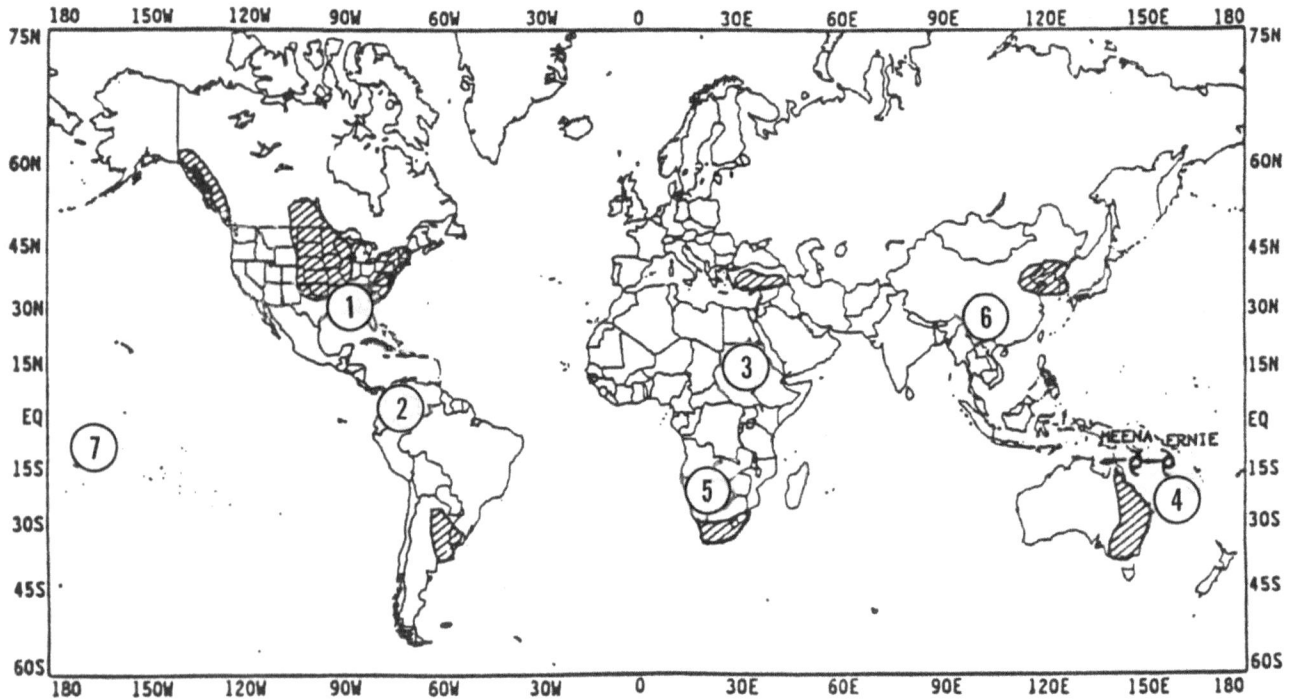

FOR THE WEEK ENDING MAY 13, 1989

Persistent Conditions (shaded)

BRITISH COLUMBIA, ALASKA: Coastal areas remain dry (12 weeks).

CENTRAL USA, S. CENTRAL CANADA: Abnormally dry weather continues (8 weeks).

E. USA: Cold, wet weather dominates (2 weeks).

ARGENTINA, URUGUAY: Dry weather returns (46 weeks).

TURKEY: Dry conditions persist (9 weeks); low temperatures end unusual warm weather (ended at 8 weeks).

S. AFRICA: Second week of little rain ends wet conditions (ends at 5 weeks).

E. ASIA: Above normal temperatures continue in Korea and NE China (12 weeks); temperatures near normal in south central Siberia.

E. AUSTRALIA: More wet weather moved into region (9 weeks).

INDIA, NEPAL: Drought; Heat stroke, water shortages, and crop damage.

BRAZIL: Deforestation; Satellite photos show 10% of Amazon destroyed.

NEPAL: Deforestation; 600 acres of trees cut per day due to a poltically caused fuel shortage.

Transient Events (numbered)

(1) TORNADOES: SE USA; Texas to Virginia, 23 dead.

(2) VOLCANO: COLUMBIA: Smoke and ash from mountain; eruption eminent; evacuation urged.

(3) THUNDERSTORMS: SUDAN; Alerts for flash flooding on Nile river.

(4) TROPICAL STORMS: Weak tropical cyclones moved over the South Pacific and Coral Sea.

(5) RHINOS THREATENED: NAMIBIA; Market for rhino's horn threatens extinction; dehorning with hacksaw by rangers to save animals.

(6) May 7: EARTHQUAKE; CHINA, Burma border, 5.6 mag., 1 killed, 91 injured, 5,300 homes destroyed.

(7) May 11: ATOMIC BOMB: FRANCE, Mururoa Atoll, 20-80 KT.

Solar Geomagnetic Data

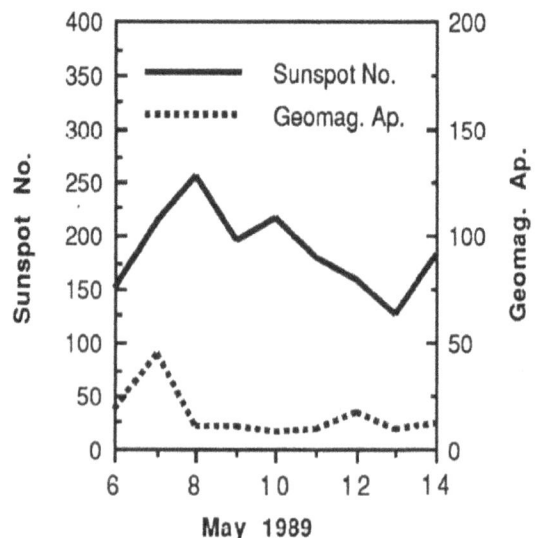

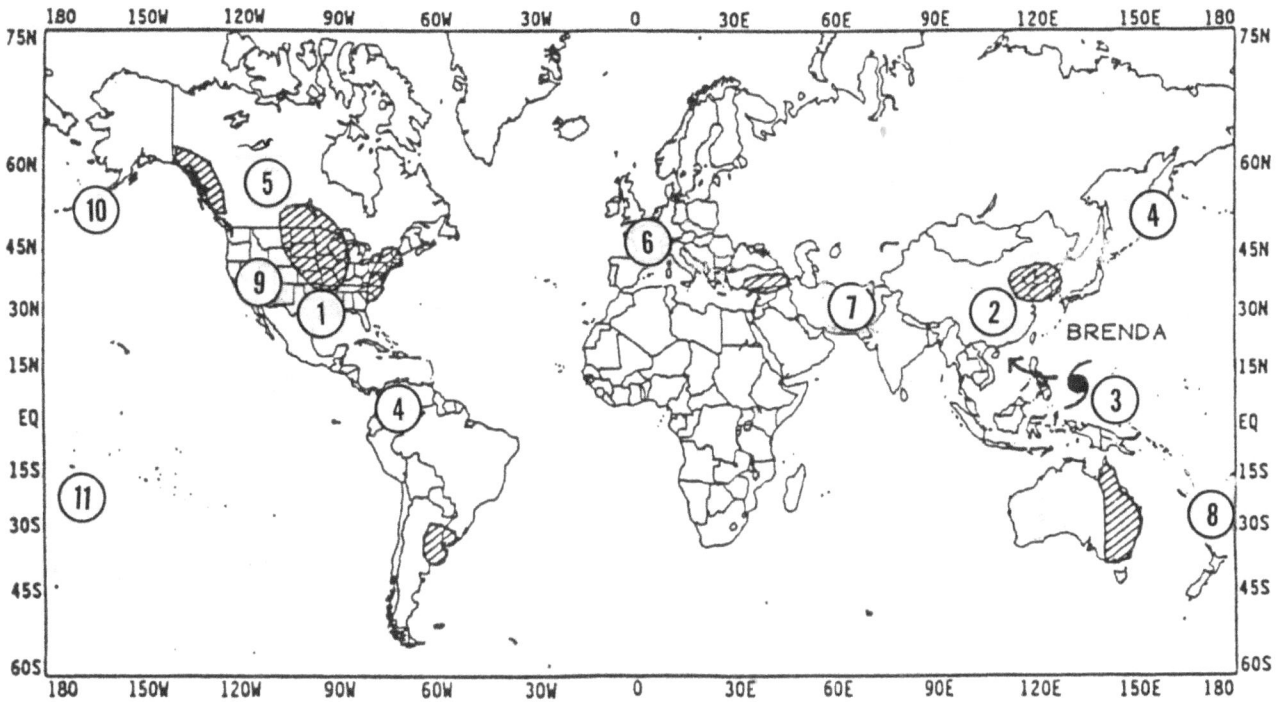

FOR THE WEEK ENDING MAY 20, 1989

Persistent Conditions (shaded)

BRITISH COLUMBIA, ALASKA: Very dry conditions persist on coast (13 weeks).

CENTRAL USA, S. CENTRAL CANADA: Light rains ease dryness (9 weeks).

E. USA: Rains continue; cold weather returns to near normal (ended at 3 weeks).

ARGENTINA, URUGUAY: Heavy rains relieve dryness in some areas (ending at 47 weeks).

TURKEY: Very dry weather continues (10 weeks).

E. ASIA: Cooler air ends warm spell (ended at 12 weeks).

E. AUSTRALIA: Heavy rains, wet weather continue (10 weeks)

INDIA, BANGLADESH: Drought; Heat kills 500 people, water shortages.

Transient Events (numbered)

(1) HEAVY RAINS: LOUSIANA, E. TEXAS; Severe flooding and 21 tornadoes reported.

(2) TORNADO: CHINA; 14 people killed, 44 injured, crops and houses destroyed.

(3) TROPICAL STORM: PHILIPPINES; Flooding caused in Manila; 2000 flee homes; moves inland over China.

(4) VOLCANOES: SOVIET FAR EAST, COLUMBIA; Populations in nearby areas prepared to move.

(5) PRAIRIE/FOREST FIRES: CANADA; 1 million acres burned.

(6) BEE ATTACK: FRANCE; 400,000 bees leave hives, swarm and kill animals.

(7) NEAR EXTINCTION: AFGHANISTAN; Only 23 Siberian cranes left after war depletes last flock.

(8) May 14: EARTHQUAKE; S. PACIFIC, Kermadec Islands, 6.6 mag.

(9) May 15: ATOMIC BOMB; USA, Nevada, <20 KT.

(10) May 19: EARTHQUAKE; ALASKA, Aleutian Islands, 6.1 mag.

(11) May 20: ATOMIC BOMB; FRANCE, Mururoa Atoll, 2 KT.

(12) May 20: FULL MOON.

Solar Geomagnetic Data

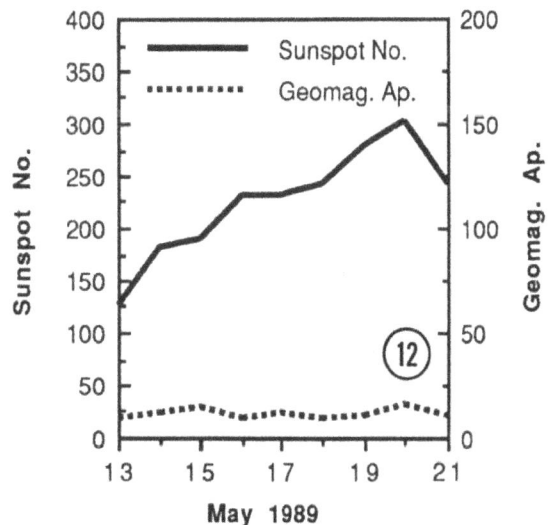

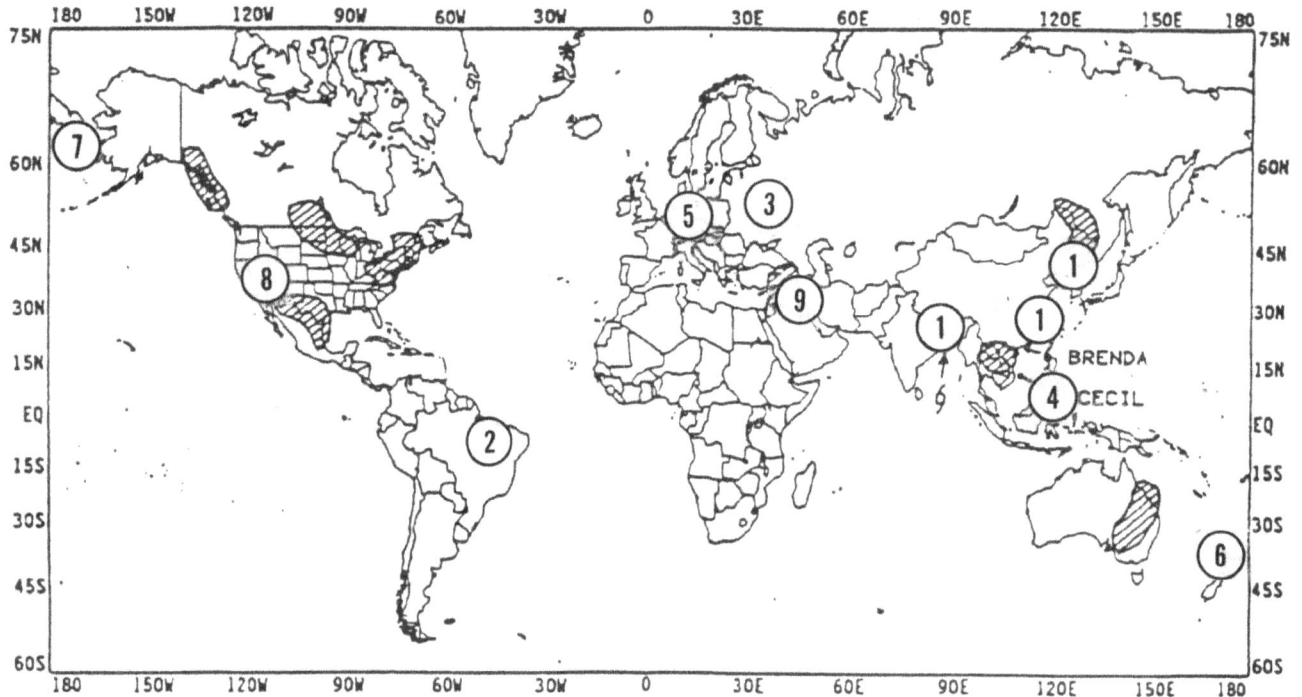

FOR THE WEEK ENDING MAY 27, 1989

Persistent Conditions (shaded)

BRITISH COLUMBIA, ALASKA: Rains return to region, dryness over (ending at 13 weeks).

N. CENTRAL USA, S. CENTRAL CANADA: Rains bring some relief, dryness persists in Corn Belt (10 weeks).

E. USA: Wet conditions persist, flooding in Ohio (4 weeks).

E. MEXICO, S. TEXAS: Heat wave develops (2 weeks).

TURKEY, SYRIA: Area remains dry (11 weeks).

MANCHURIA, SE USSR: Insufficient rains develop dryness (4 weeks).

E. AUSTRALIA: "Big Wet" continues; rains slightly less.

ENGLAND: Drought; Water reserves fall, temperatures soar.

INDIA: Drought; Heat wave continues, water shortages; tropical cyclone may bring relief.

Transient Events (numbered)

(1) TROPICAL STORMS: Southeastern China, Northern Vietnam, and India experience rains and high winds.

(2) HEAVY RAINS: BRAZIL; Floods and mudslides kill 54, thousands homeless.

(3) GIANTISM: USSR; Some plants in Chernobyl area growing to huge sizes.

(4) CROCODILE ATTACK: PHILIPPINES; Rare reptile eats 12 people, attacks animals.

(5) PIGEON PLAGUE: W. GERMANY; Prize offered for pigeon birth control pill.

(6) May 23: EARTHQUAKE; NEW ZEALAND, Macquarie Islands, 6.0 mag.

(7) May 24: EARTHQUAKE; USSR, Komandorsky Isl., 6.1 mag.

(8) May 26: ATOMIC BOMB; U.S.A., Nevada, <20 KT.

(9) May 27: EARTHQUAKE; IRAN; 5.7 mag., Houses damaged.

Solar Geomagnetic Data

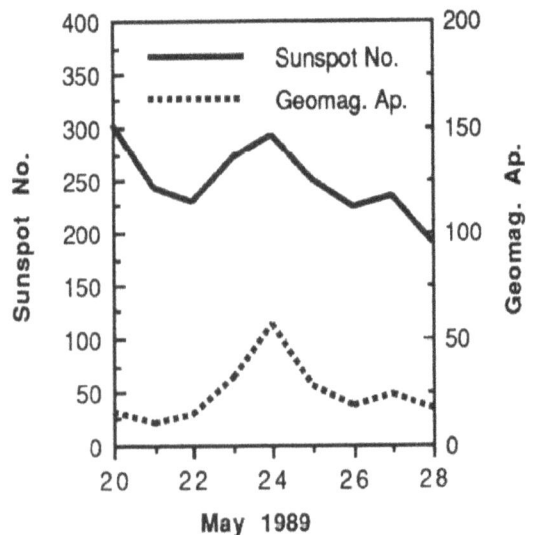

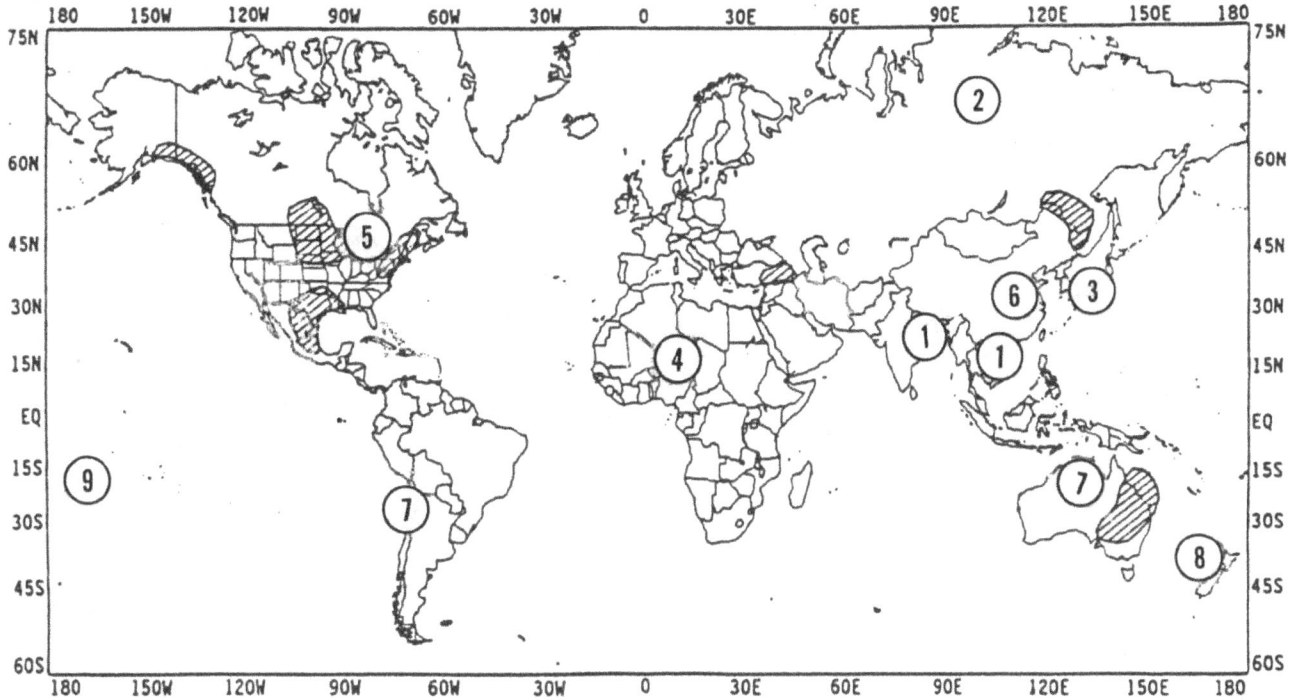

FOR THE WEEK ENDING JUNE 3, 1989

Persistant Conditions (shaded)

BRITISH COLUMBIA, ALASKA: More rain definitely ends dry spell (ends at 14 weeks).

N. CENTRAL USA, S. CENTRAL CANADA: Some rain, dryness continues (11 weeks).

E. USA: Wetness diminishes, central New York receives showers (5 weeks).

E. MEXICO, S. TEXAS: Heat wave continues (3 weeks).

TURKEY: Area still dry (12 weeks).

MANCHURIA, SE USSR: Rains ease dryness in Manchuria, other areas still subnormal (5 weeks).

E. AUSTRALIA: Drier conditions bring relief from wetness (12 weeks).

Transient Events (numbered)

(1) TROPICAL STORM AFTERMATH: INDIA: Heat wave and drought broken, tidal waves and tornadoes kill over 200 people, thousands homeless; VIETNAM: 100,000 homeless, 140 killed, crops and buildings damaged; MEXICO, Storm forms off Baja.

(2) FOREST FIRES: SIBERIA; Hot and dry winds fanned fires.

(3) ATOMIC FIRE: JAPAN; Spontaneous combustion of uranium waste occurred at atomic energy plant; 2 1/2 hours to extinguish.

(4) LOCUSTS: NIGER; New swarms invade area.

(5) EEL MENACE: USA, CANADA; Non-native population explodes in Great Lakes, threatens native trout.

(6) ENDANGERED SPECIES: CHINA; Bear, python and leopard parts served in restaurants despite government protection; police raid the restaurants.

(7) EARTHQUAKES; AUSTRALIA, 5.7 mag.; CHILE, 5.7 mag.

(8) May 31: EARTHQUAKE; NEW ZEALAND; 6.4 mag.

(9) June 3: ATOMIC BOMB; FRANCE, Mururoa Atoll, 10-40 KT.

(10) June 3: NEW MOON.

Solar Geomagnetic Data

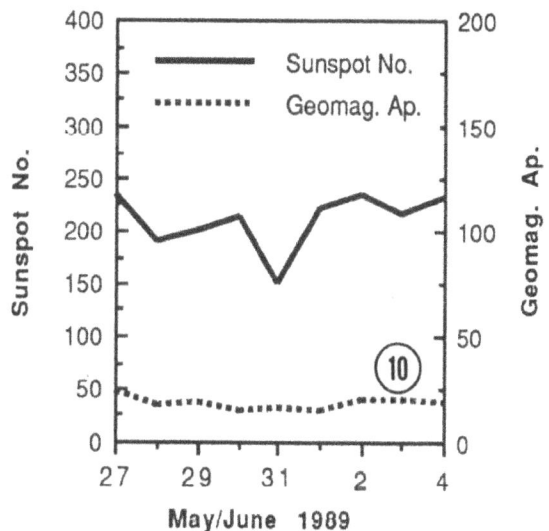

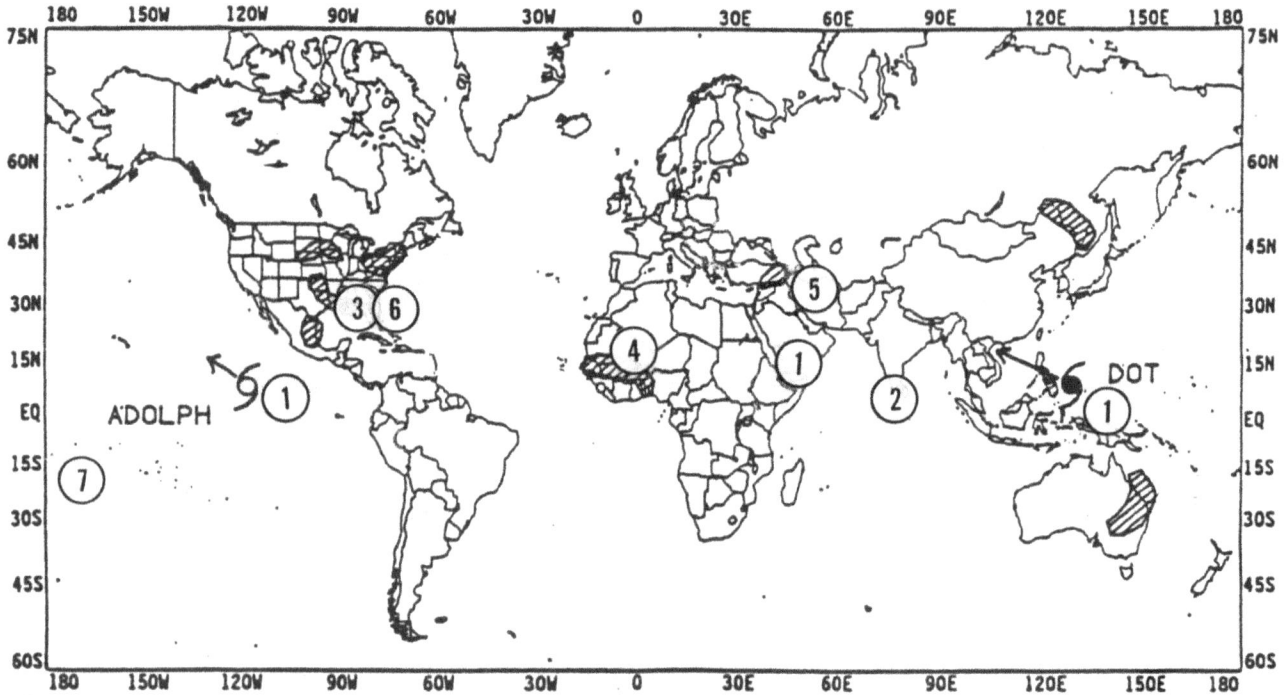

FOR THE WEEK ENDING JUNE 10, 1989

Persistent Conditions (shaded)

N. CENTRAL USA: Increased rainfall limits dryness to isolated areas (12 weeks).

E. U.S.A.: Wet conditions continue (6 weeks).

S. GREAT PLAINS, GULF COAST USA: Torrential rains saturate area (4 weeks).

E. MEXICO, S. TEXAS: Above normal temperatures continue (4 weeks).

SAHELIAN W. AFRICA: High temperatures continue (6 weeks).

TURKEY: Some rainfall, dryness persists (13 weeks).

MANCHURIA, SE USSR: Rains continue to ease dryness(ended at 5 weeks).

E. AUSTRALIA: Rains return, "Big Wet" continues (13 weeks).

Transient Events (numbered)

(1) TROPICAL STORMS: Storm travels from Philippines to south China; weak cyclone brings rain to Oman, South Yemen, Southern Arabia; storm off Mexico.

(2) MONSOONS: INDIA, SRI LANKA: Southwest monsoon from Africa brings rains and cools India; floods Sri Lanka.

(3) TORNADOES: E. USA; Large hail and flooding; 2 people killed in Louisiana.

(4) LOCUSTS: MALI, NIGER: New swarms spotted, plague feared.

(5) DEER THREATENED: IRAN, IRAQ: War has decimated the populations of herds.

(6) ROACHES RELEASED: FLORIDA; Imported, mouse -size hissing cockroaches were accidentally released into the wild, fear they may multiply.

(7) June 10: ATOMIC BOMB; FRANCE; Fangataufa Atoll, 20-80 KT.

Solar Geomagnetic Data

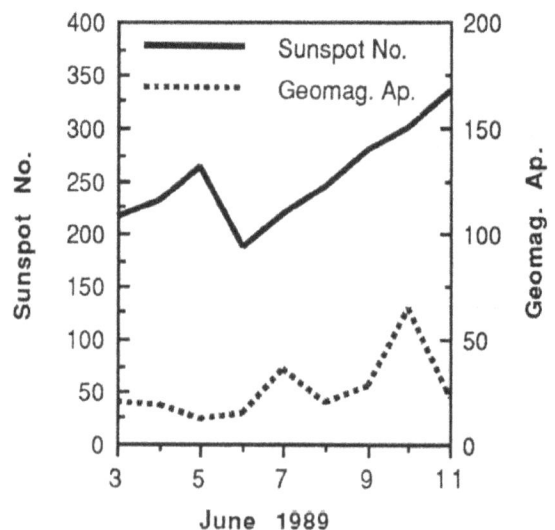

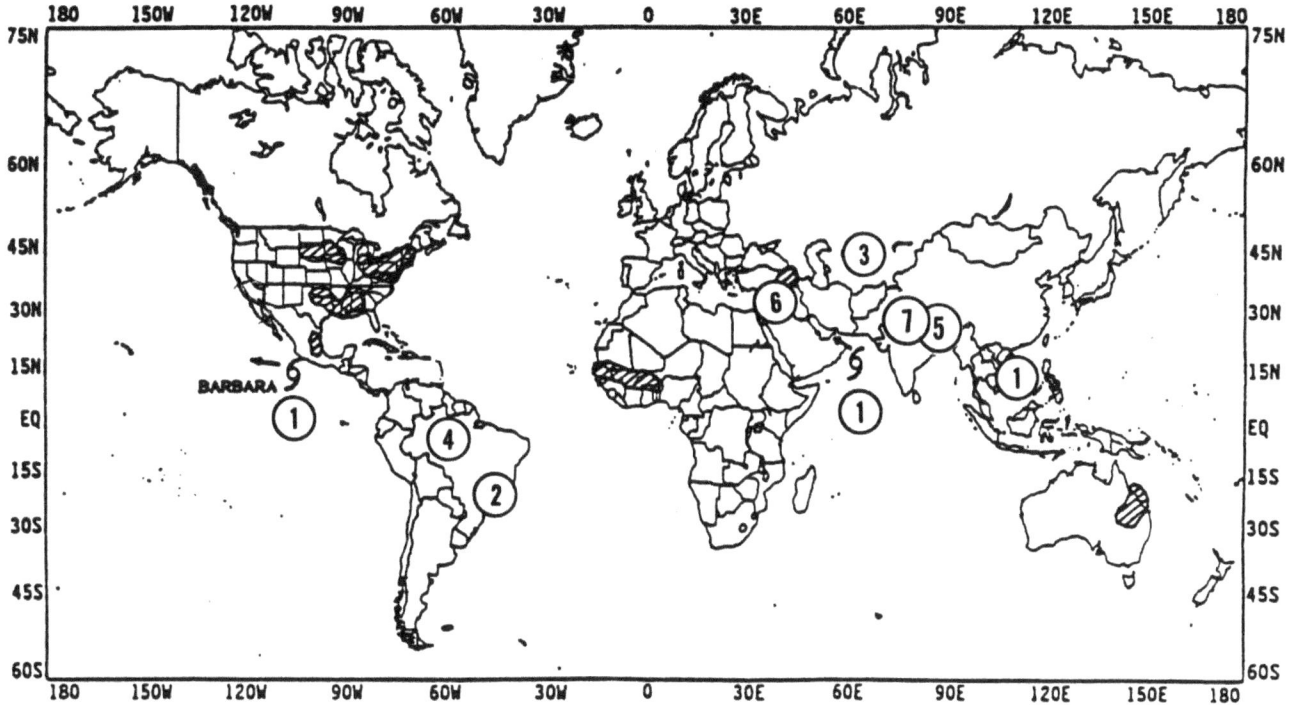

FOR THE WEEK ENDING JUNE 17, 1989

Persistent Events (shaded)

N. CENTRAL USA: Widespread rainfall, but long-term precipitation deficits remain (13 weeks).

E. USA: Rains continue to soak region (7 weeks).

S. GREAT PLAINS, GULF COAST USA: Excessive rains continue (5 weeks).

E. MEXICO: Heat wave diminishes, still hot (5 weeks).

SAHELIAN W. AFRICA: Temperatures cool to normal in most areas (7 weeks).

TURKEY: Moderate rainfall ends dryness (13 weeks).

E. AUSTRALIA: Normal dryness returns to wet area (14 weeks).

NEW ZEALAND: Past 12 months are warmest on record; blamed on La Nina warm ocean water.

Transient Events (numbered)

(1) TROPICAL STORMS: Typhoon tears through Vietnam; second cyclone forms in the Arabia Sea; third off Central America.

(2) HEAVY RAINS: BRAZIL; Mudslides and floods afflict Rio de Janeiro.

(3) GLACIER WARNING: ASIAN USSR; Glacier blocks river and threatens flooding.

(4) DEFORESTATION: BRAZIL; Governor of Amazonas plans to distribute 4,000 power saws to encourage "development" of threatened Amazon rainforest.

(5) BUTTERFLY PLAGUE: BANGLADESH; Clouds of Burmese butterflies devour crops.

(6) OPIUM/HASH CROP: LEBANON; Siberian cold wave decreases harvest by 60%.

(7) June 12: EARTHQUAKE; BANGLADESH; 6.0 mag.

Solar Geomagnetic Data

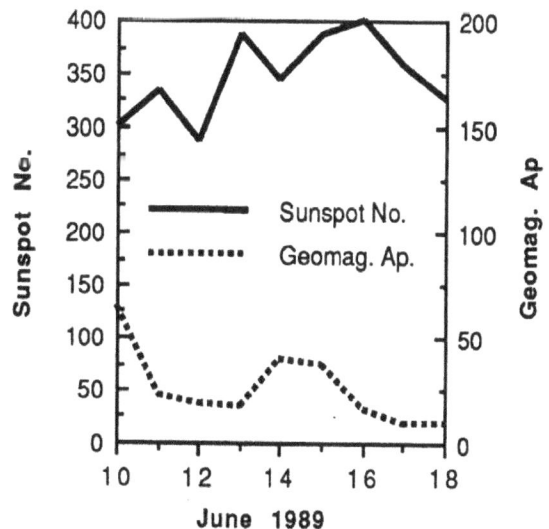

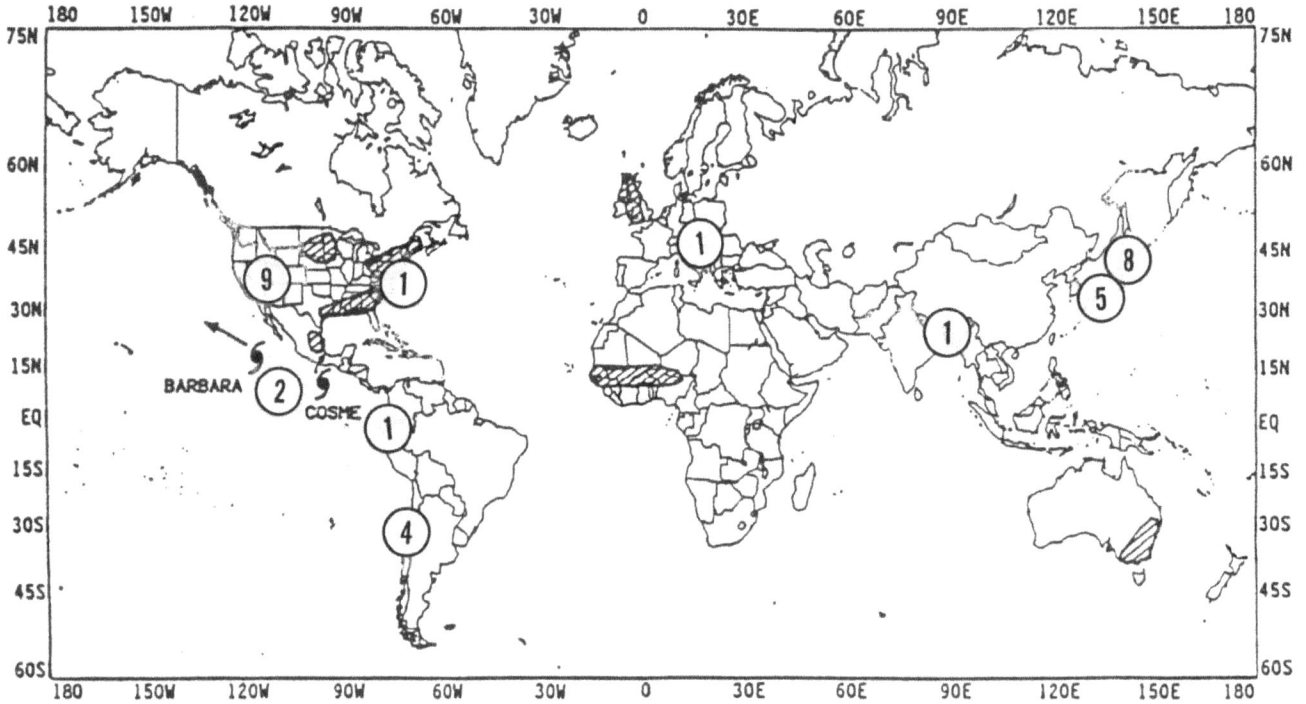

FOR THE WEEK ENDING JUNE 24, 1989

Persistent Conditions (shaded)

N. CENTRAL USA: Rainfall has not ended precipitation deficits (14 weeks).

NE USA: More heavy rain (8 weeks).

GULF COAST, USA: Wet weather continues; flooding (6 weeks).

E. MEXICO: Heat wave ends (ended at 5 weeks).

BRITISH ISLES: Dry conditions continue (4 weeks).

SAHELIAN W. AFRICA: Hot conditions over (ended at 7 weeks).

SE AUSTRALIA: Rains return wet conditions (15 weeks).

Transient Events (numbered)

(1) FLOODS: ECUADOR, YUGOSLAVIA, ITALY, BANG-LADESH, E. USA.

(2) TROPICAL STORMS: Hurricane *Cosme* struck Mexican coast; 10 inches of rainfall and flooding reported: *Barbara* lost force over cool waters of Baja.

(3) WILDFIRE: SOVIET FAR EAST; 3,000 acres destroyed on island of Sakhalin.

(4) VOLCANO: CHILE; 5,000 people evacuated when toxic gas spewed from peak.

(5) DOG ODYSSEY: JAPAN; After two years, 42 miles, six-year old dog finds former owner.

(6) June 19: FULL MOON.

(7) June 21: SUMMER SOLSTICE: Sun reached northernmost position of year; longest days in N. Hemisphere.

(8) June 21: EARTHQUAKE; JAPAN; 6.8 mag.

(9) June 22: ATOMIC BOMB; USA, Nevada, 20-150 KT.

Solar Geomagnetic Data

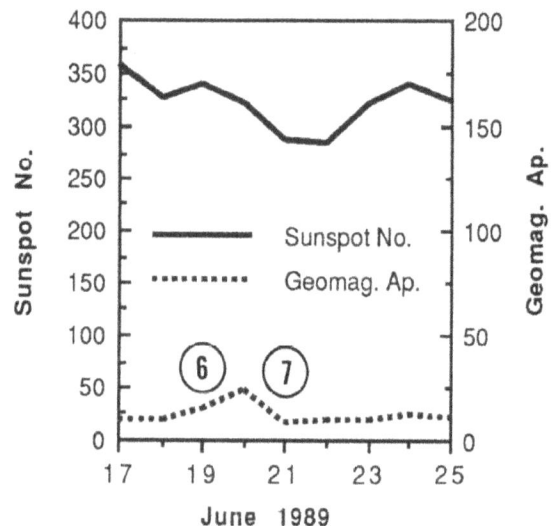

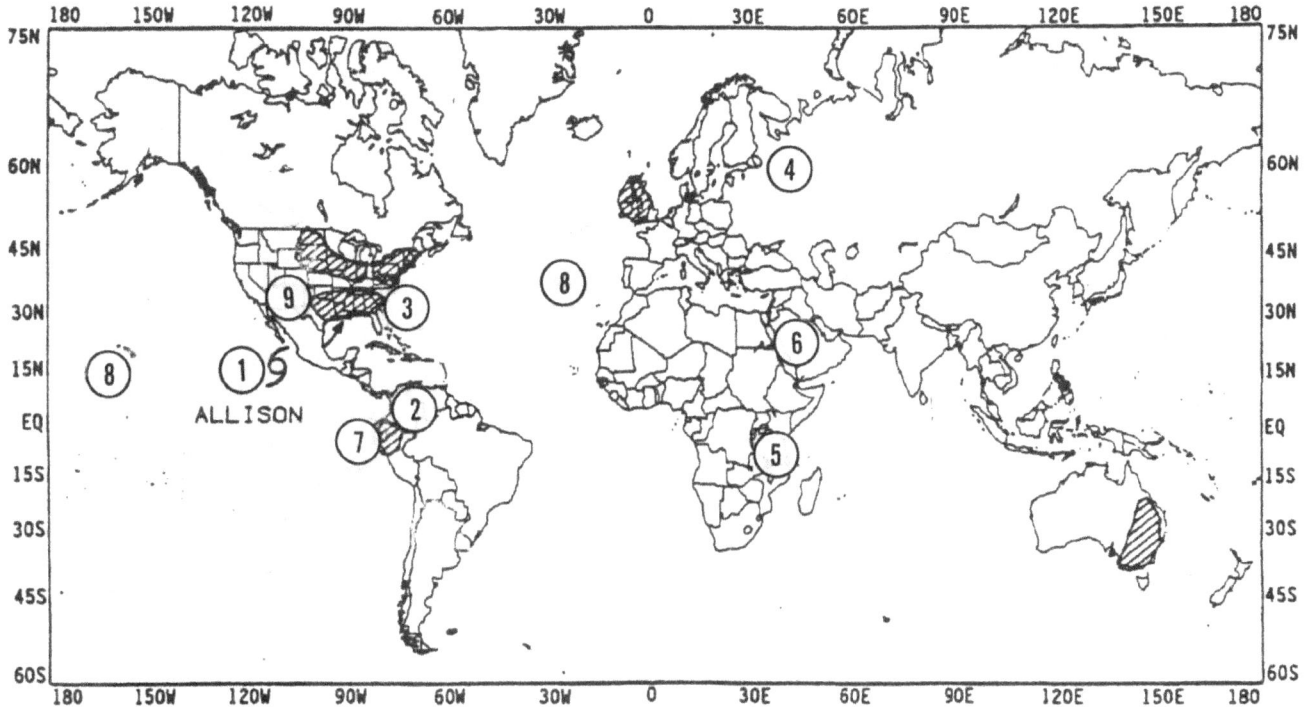

FOR THE WEEK ENDING JULY 1, 1989

Persistent Conditions (shaded)

N. CENTRAL USA: Rainfall continues, but dry deficits remain (15 weeks).

NE USA: Dry conditions relieve wet weather (ended at 8 weeks).

GULF COAST, USA: Rains continue (7 weeks).

ECUADOR: Rains continue (2 weeks).

BRITISH ISLES: Rains relieve dryness (ended at 4 weeks).

SE AUSTRALIA: Inland areas drier; rains continue on coast (16 weeks).

ARGENTINA, URUGUAY: Drought continues into "normal" dry season. Electric and water shortages.

Transient Events (numbered)

(1) TROPICAL STORM: *Allison* traveled across Mexico into Houston and Louisiana causing tornadoes and flooding.

(2) VOLCANO: COLUMBIA; Ashes spewed and seismic activity increased in killer volcano.

(3) OIL SPILLS: GALVESTON BAY, DELAWARE BAY, NEWPORT. RI; Oil tanker spills marred these three areas.

(4) UFO SIGHTINGS: USSR; Number of sightings near Vologda region.

(5) CROCODILES: TANZANIA; Government is killing crocodiles because over- population threatens fish.

(6) MONKEY RAID: SAUDI ARABIA; Restaurant attacked by troop of mountain monkeys who took food from customer's plates and ran.

(7) June 25: EARTHQUAKE; ECUADOR, Coast, 6.0 mag., slight damage.

(8) June 26: EARTHQUAKES: HAWAII; 6.0 mag.; strongest in years near volcano, 5 people injured, homes damaged, small tsunami; AZORES ISLANDS; 5.8 mag., people injured, some damage.

(9) June 27: ATOMIC BOMB: USA; NEVADA, 20-150 KT.

Solar Geomagnetic Data

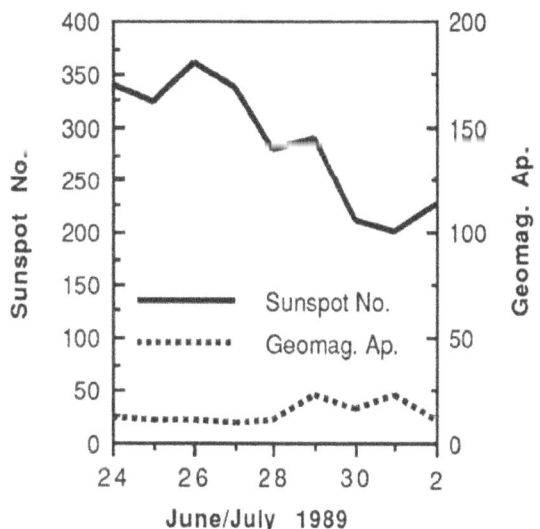

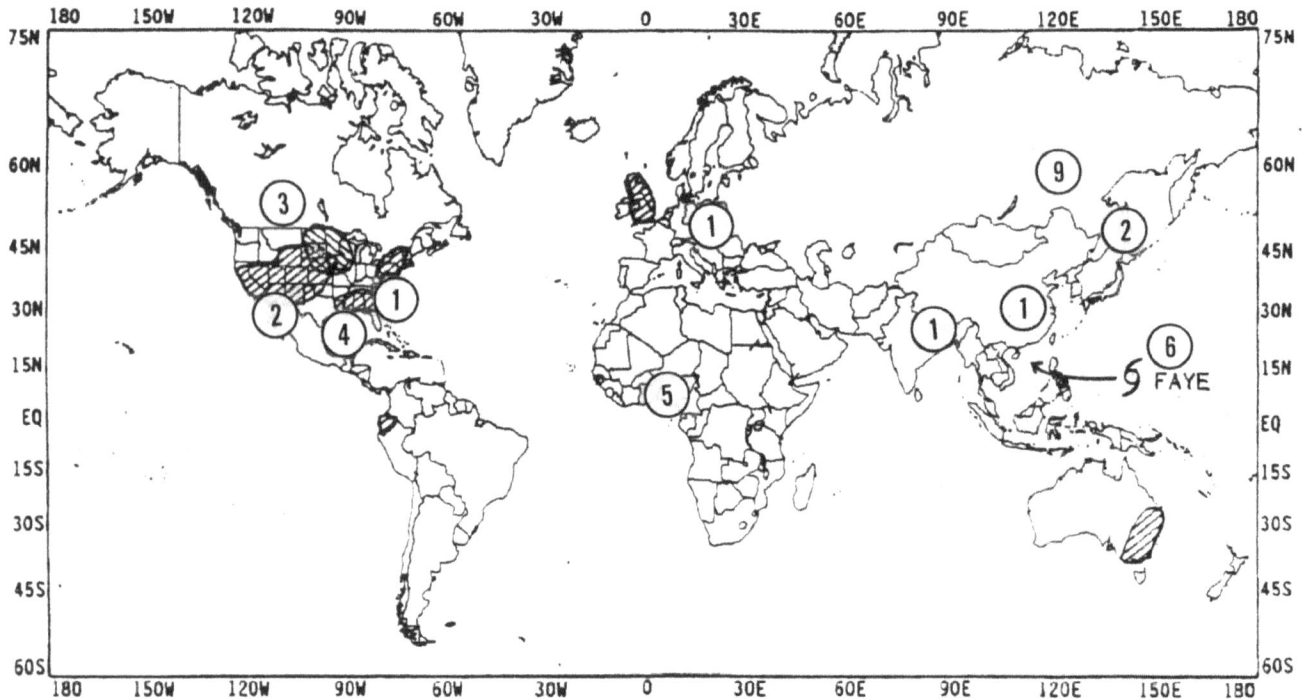

FOR THE WEEK ENDING JULY 8, 1989

Persistent Conditions (shaded)

W. CENTRAL USA: Heat wave occurs (2 weeks).

N. CENTRAL USA: Long-term dryness persists (16 weeks).

NE USA: Rains continue (9 weeks).

GULF COAST: Wet weather persists (8 weeks).

ECUADOR: Heavy rains continue (3 weeks).

BRITISH ISLES: Dryness ends (ended at 4 weeks).

SE AUSTRALIA: Wetness diminishes (ends at 16 weeks).

HAITI: Deforestation; Once lush island will become a man-made desert in 20 years if trees not saved; only 1% of island remains wooded.

Transient Events (numbered)

(1) FLOODS: CHINA, INDIA, BANGLADESH, CENTRAL EUROPE, SE USA.

(2) WILDFIRES: SAKHALIN ISLAND, USSR: 20% of island's surface burned; S. CALIFORNIA: Homes destroyed in hills.

(3) TORNADOES: CANADIAN PRAIRIES; Building and crop damage.

(4) ALGAE: GULF OF MEXICO; Sargassum piles up on beaches; unusual currents may have brought it ashore from mid-Atlantic.

(5) HYACINTH: NIGERIA; Plant clogs rivers; spreading to all coastal states.

(6) TROPICAL STORM: *Faye* heads toward China.

(7) July 3: NEW MOON.

(8) July 4: EARTH AT APHELION, farthest from Sun.

(9) July 8: ATOMIC BOMB; USSR, Semipalatinsk, magnitude not available.

Solar Geomagnetic Data

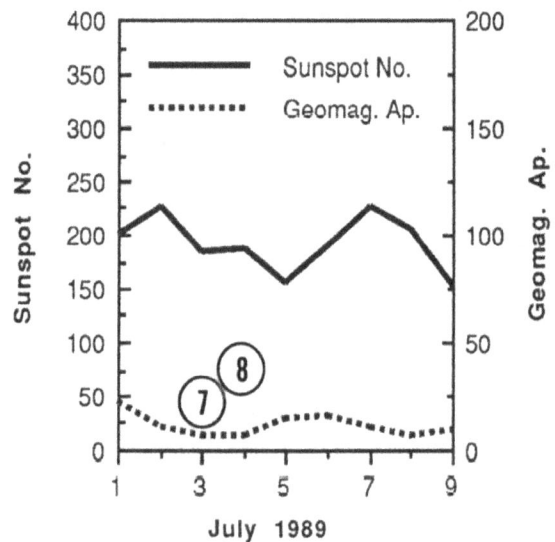

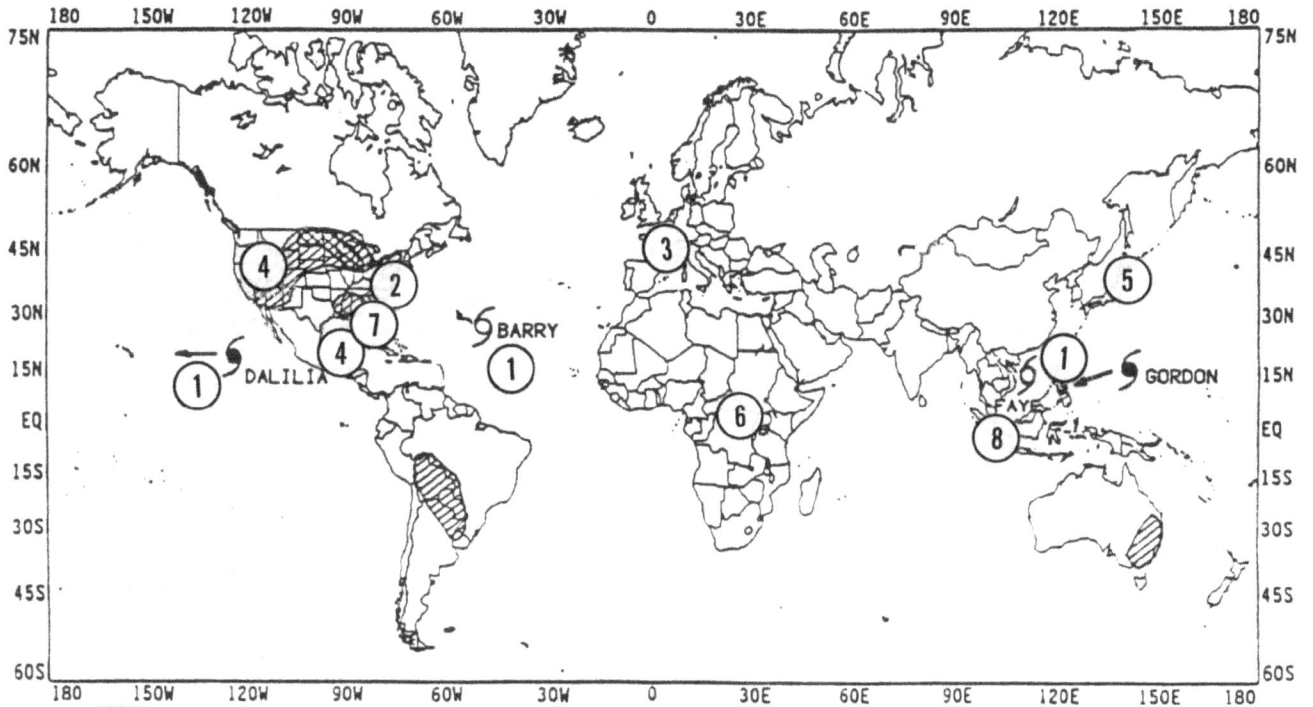

FOR THE WEEK ENDING JULY 15, 1989

Persistent Conditions (shaded)
W. CENTRAL USA: Heat wave eases (3 weeks).
N. CENTRAL USA: Rains occur, but long tern dryness remains (17 weeks).
NE USA: Rains diminish (10 weeks).
GULF COAST USA: Wetness eases (9 weeks).
S. CENTRAL AMERICA: Abnormally cold weather spreads (2 weeks).
SE AUSTRALIA: "Big Wet" ends (ended at 16 weeks).

Transient Events (numbered)
(1) TROPICAL STORMS: Typhoon *Faye* inundates China, floods kill over 800; three other storms remain at sea in the Pacific and Atlantic.
(2) TORNADOES: New York, New Jersey and Connecticut hit during violent thunderstorms.
(3) HEAVY RAINS: FRANCE; Violent thunderstorms cause crop damage.
(4) WILDFIRES: YUCATAN, 2.5 thousand sub-tropical forest acres burned; W. USA, Fires rage, but higher humidity helps firefighters.
(5) VOLCANO: JAPAN; Undersea volcano erupts two miles from mainland; preceded by swarms of earthquakes; evacuations planned.
(6) HIPPO ATTACK: ZAIRE; Student killed when hippo chased and overturned canoe; angry hippos had just escaped poachers.

(7) POISONOUS TOADS: FLORIDA; Non-native toads escaped and breeding in wilds; poison glands can kill pets.
(8) July 14: EARTHQUAKES; INDONESIA, Timor, 6.4 mag., people injured, some damage.

Solar Geomagnetic Data

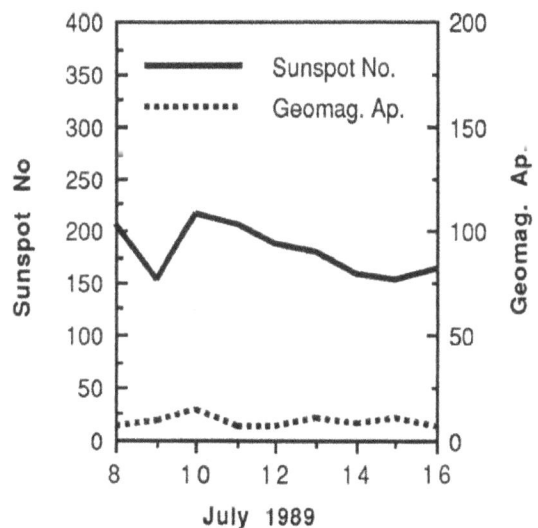

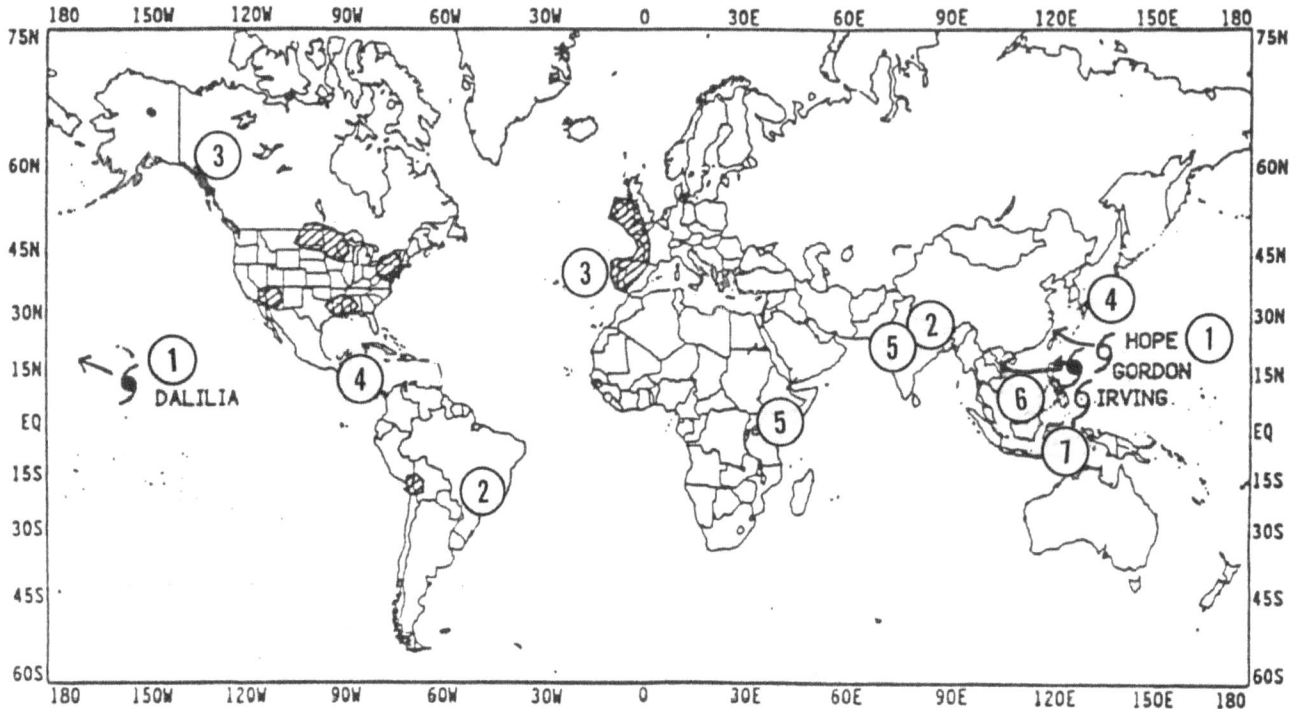

FOR THE WEEK ENDING JULY 22, 1989

Persistent Conditions (shaded)

W. USA: Heat wave ends (ended at 3 weeks).

N. CENTRAL USA: Some rains, long-term dryness continues (18 weeks).

NE USA: Wetness persists (11 weeks).

GULF COAST USA: Rains diminish (ending at 9 weeks).

CENTRAL S. AMERICA: Cold spell over (ended at 2 weeks).

W. EUROPE: Heat wave occurs (2 weeks).

Transient Events (numbered)

(1) TROPICAL STORMS: Typhoon *Gordon* swept northern Philippines, killing 32, then Hong Kong, and southern China, killing more. Two more storms form in same area, headed towards land. *Dahlia* caused flooding in Hawaii.

(2) FLOODING: BANGLADESH; INDIA: Many killed and homeless after monsoon rains; BRAZIL: Heavy rains leave 410,000 homeless.

(3) WILDFIRES: PORTUGAL, Worst fires in 20 years; CANADA, Fires burn across Yukon.

(4) VOLCANOES: GUATEMALA, Santiaguito erupts; JAPAN, Mount Aso erupts; undersea volcano still sending shocks.

(5) ELEPHANTS: INDIA, 16 rogue elephants rampage through village; KENYA, President burns 12 tons of confiscated poacher's ivory. Only 17,000 elephants left in Kenya.

(6) EELS: HONG KONG, Moray eels attack swimmers; weather changes caused eel migration to bay.

(7) July 22: EARTHQUAKE; INDONESIA, Halmahera, 6.4 mag.; one killed, building damaged.

(8) July 18: FULL MOON.

Solar Geomagnetic Data

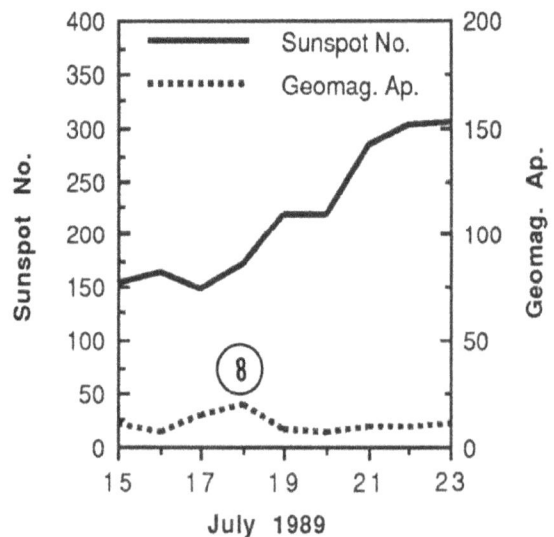

July 1989

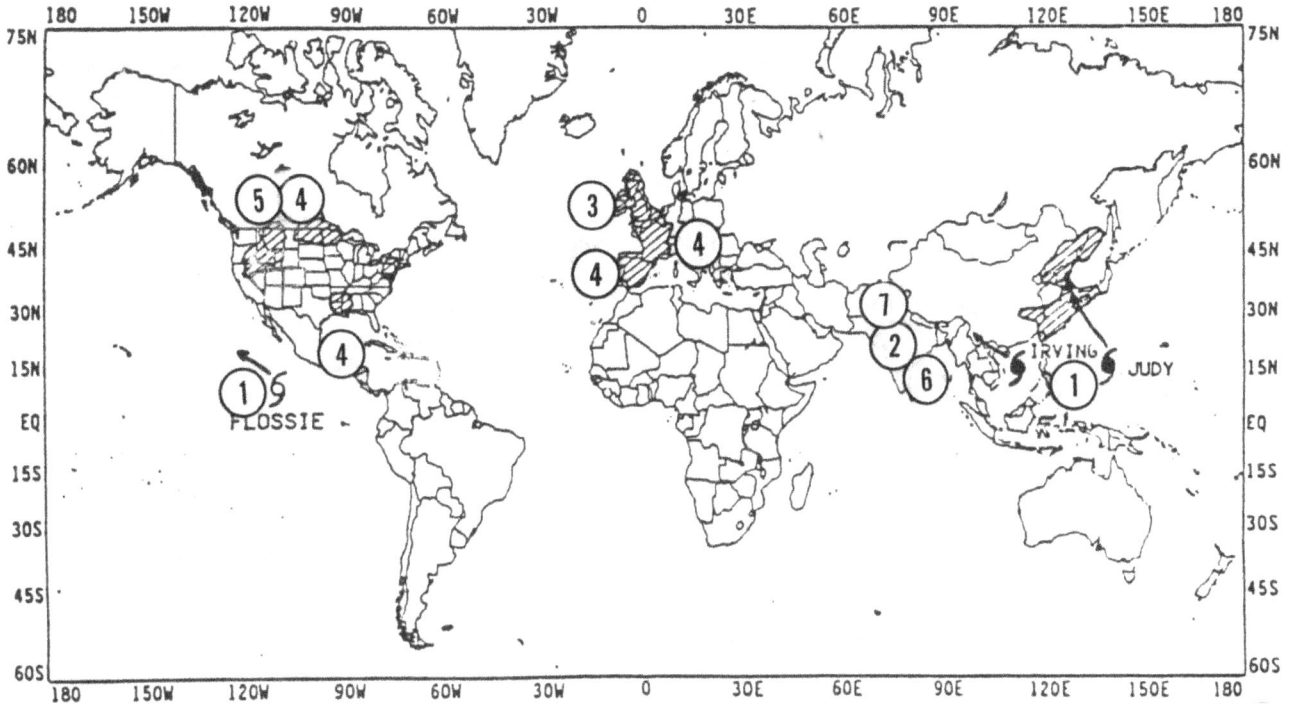

FOR THE WEEK ENDING JULY 29, 1989

Persistent Conditions (shaded)
NW USA: Heat wave occurs (2 weeks).
N. CENTRAL USA: Some rain, but dry deficits continue (19 weeks).
NE USA: Wetness diminishes (ending at 11 weeks).
GULF COAST, USA: Scattered rain; wetness diminishes (ending at 9 weeks).
W. EUROPE: Hot weather remains (3 weeks).

Transient Events (numbered)
(1) TROPICAL STORMS: *Irving* moved inland over Vietnam, killing 31; *Judy* lashed southern Japan with 115 mph winds, killing 2; flooding in Taiwan, Korea, and areas of China. Another storm in Pacific remains at sea.
(2) HEAVY RAINS: INDIA, 800 people feared dead, 1000s missing from floods and mudslides from monsoon rains.
(3) AIR POLLUTION: ENGLAND; Continued drought causes high temperatures, water shortages, and air twice as bad as safe level.
(4) WILDFIRES: W. CANADA: Rain brings fires under control; 27,000 evacuated; YUCATAN PENINSULA: 265,000 acres destroyed; SPAIN: Fires threaten nuke plant; GREECE: U.S. military base threatened.
(5) ELK POLLUTION: ALBERTA, Genetically hybrid elk escapes game farm, may breed with native elk population; animal lacks disease resistance.

(6) TREE FUNERAL: INDIA, 1,000 year old tree cremated with full religious rites; was uprooted in storm.
(7) EARTHQUAKE: PAKISTAN, 5.6 mag.

Solar Geomagnetic Data

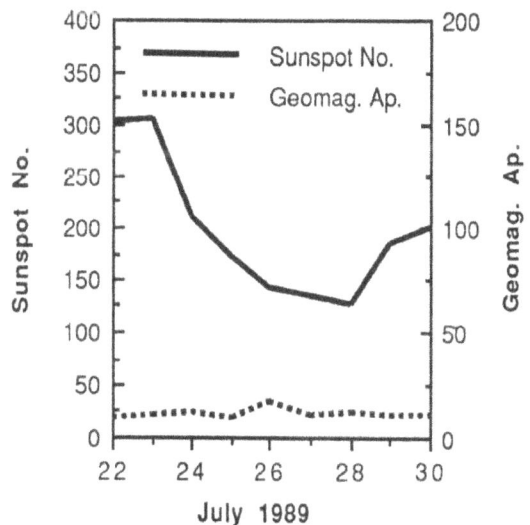

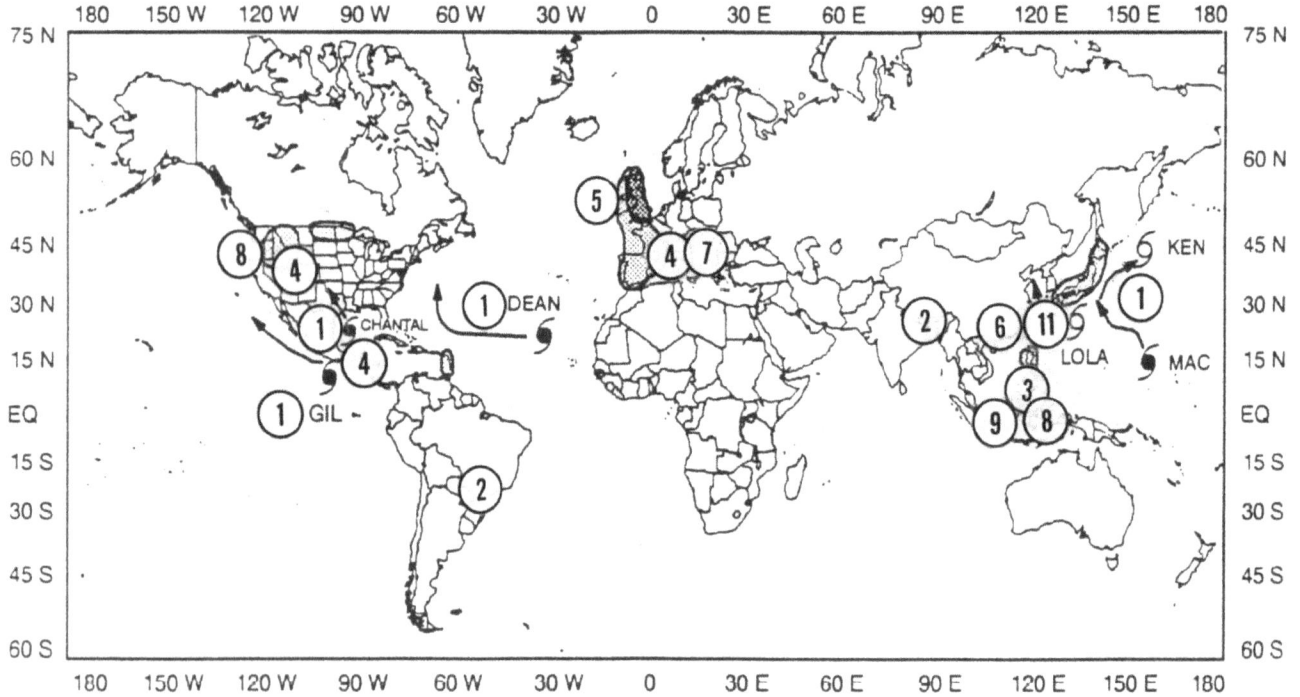

FOR THE WEEK ENDING AUGUST 5, 1989

Persistent Conditions (shaded)

NW USA: Extreme heat subsides (ending at 2 weeks).

N. CENTRAL USA: Rains diminish dryness (ended at 19 weeks).

CARIBBEAN ISLANDS: Lack of tropical rains creates dryness (8 weeks).

BRITISH ISLES: Drought continues (5 weeks).

W. EUROPE: Heat wave continues (4 weeks).

Transient Events (numbered)

(1) TROPICAL STORMS: *Ken* and *Lola* thrashed Japan; Hurricane *Chantal* caused flooding and tornado near Houston; *Dean* skirted the Caribbean; *Gil* heads north; and *Mac* forms in Pacific headed for Japan.

(2) HEAVY RAINS: INDIA; BANGLADESH: 1,400 people killed, millions flee homes; BRAZIL: 36 people killed in flooding.

(3) TORNADO: PHILIPPINES, 5 people killed when tornado hits Manila suburb.

(4) WILDFIRES: W. USA: 215,000 acres burned in five states; S. FRANCE: 93 mph Alpine winds fanned fires along Med.; MEXICO: Rains helped to control fires in Yucatan Peninsula.

(5) SUNSHINE: ENGLAND: Drought has brought sunniest number of days on record: 839 hours for three months.

(6) COOL WINDS: SE CHINA: Hong Kong experienced lowest temperatures since records kept in 1884:

overnight low of 71 degrees.

(7) ALGAE: ADRIATIC COAST: Slimy algae killing marine life and tourist trade: $76 paid to tourists who return to Venice.

(8) July 31: EARTHQUAKE; INDONESIA, Flores Isl., 6.3 mag., slight damage.

(9) Aug. 1: EARTHQUAKE; INDONESIA, West Irian, 5.9 mag., 90 killed, landslides bury village and block river.

(10) Aug. 1: NEW MOON.

(11) Aug. 3: EARTHQUAKE; TAIWAN, 6.3 mag.

Solar Geomagnetic Data

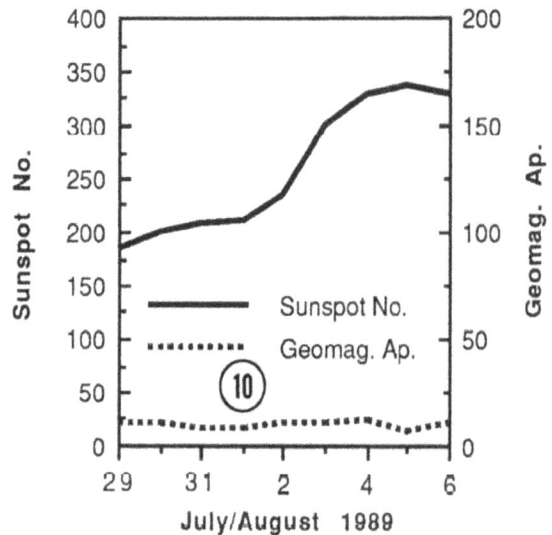

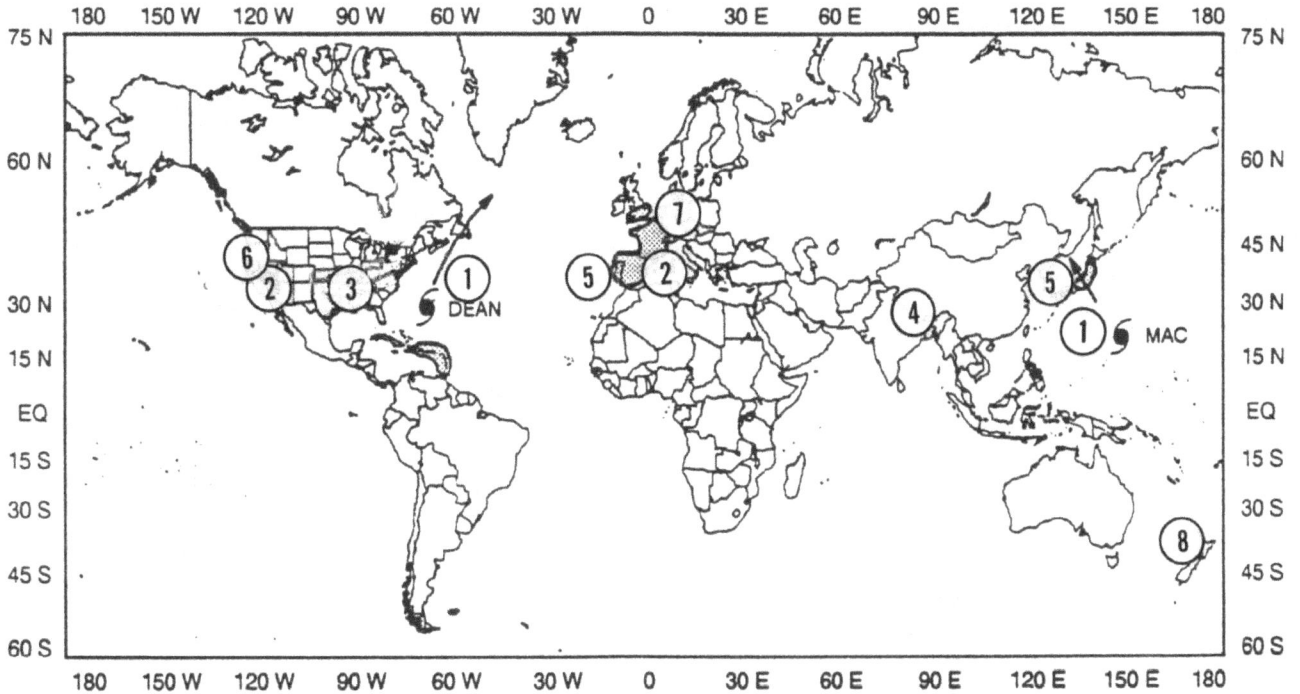

FOR THE WEEK ENDING AUGUST 12, 1989

Persistent Conditions (shaded)
CARIBBEAN ISLANDS: Dryness continues (9 weeks).
BRITISH ISLES: Drought continues (6 weeks).
W. EUROPE: Heat wave continues (5 weeks).
JAPAN: Heavy rains continue (2 weeks).

Transient Events (numbered)
(1) TROPICAL STORMS: Typhoon *Mac* hit Japan , killing five and flooding 3,000 homes; Hurricane *Dean* missed Bermuda, but hit Newfoundland, losing force.
(2) HEAVY RAINS: S. FRANCE, SPAIN: Torrential rains cause flooding, cut power; S. CALIFORNIA, W. ARIZONA: Unusual thunderstorms dump rain on arid Southwest; flooding and high winds; area received 500% of precipitation.
(3) COOL SPELL: S. & E. USA: Record-breaking low temperatures occur.
(4) FLOODS: INDIA, BANGLADESH: Monsoon rains continue floods and mudslides.
(5) HEAT WAVES: S. KOREA: 40 people drown trying to escape heat; S. SPAIN: Worst heat wave since 1922.
(6) TORNADO: OREGON: Rare tornado caused widespread damage.
(7) BEAVER RESCUE: NETHERLANDS: Bank flies reposessed beavers to Uruguay to replenish populations; originally they were to be to be killed for fur.
(8) Aug. 8: EARTHQUAKE, NEW ZEALAND, 6.3 mag., damage on both islands.

Solar Geomagnetic Data

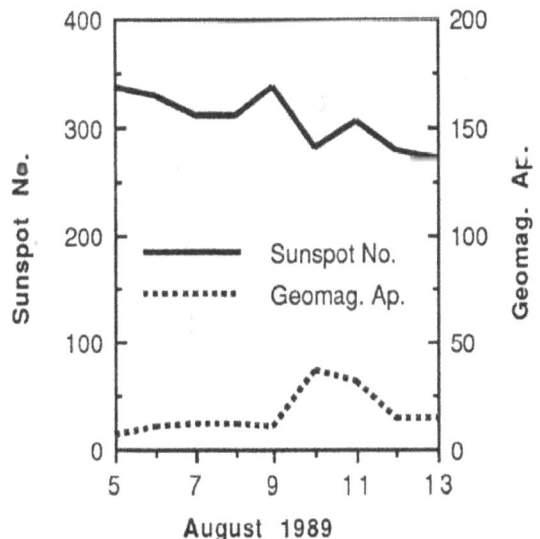

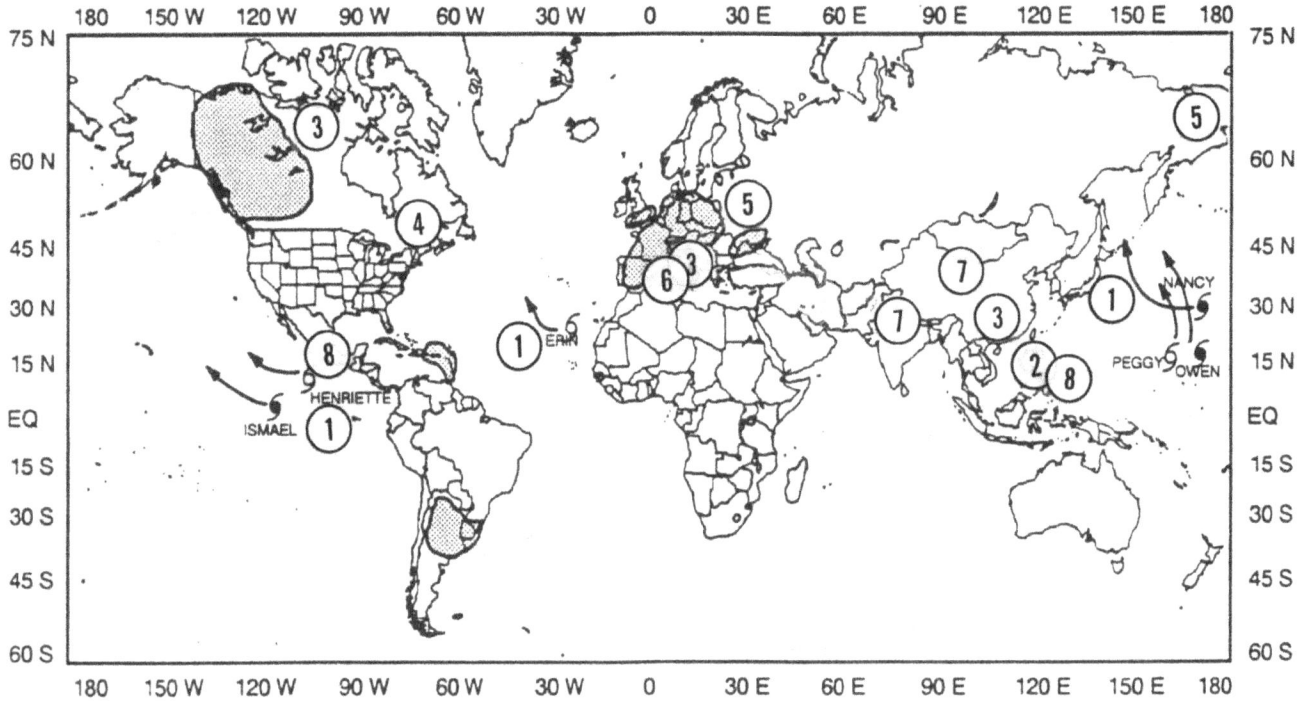

FOR THE WEEK ENDING AUGUST 19, 1989

Persistent Conditions (shaded)

W. CANADA: Warm spell persists (3 weeks).

CARIBBEAN ISLANDS: Some islands receive rains; others still dry (9 weeks).

N. ARGENTINA, URUGUAY, S. BRAZIL: Abnormally warm winter weather prevails (2 weeks).

BRITISH ISLES: Rains ease drought (ending at 7 weeks).

EUROPE: Heat wave spreads (5 weeks).

BULGARIA, ROMANIA, MOLDAVIAN, UKRAINIAN SSR: Dry conditions develop (7 weeks).

Transient Events (numbered)

(1) TROPICAL STORMS: Japan hit by Typhoons *Nancy* and *Owen*, *Peggy* goes out to sea; Hurricane *Ismael* skirts Mexican coast and moves with *Henriette* out to the Pacific; *Erin* moves into Atlantic.

(2) HEAVY RAINS: PHILIPPINES: Flooding, landslides, and rural destruction.

(3) LIGHTNING: ITALY: 2 children killed during procession; CHINA: Huge fire and explosions caused by strike on oil depot, 5 killed; CANADIAN ARCTIC: 30 caribou found dead on tundra, evidence of lightning strike nearby.

(4) SOLAR STORM: Strongest in series of sun explosions sent out stream of charged particles; may have caused power failure in Quebec.

(5) NUCLEAR CONTAMINATION: CHERNOBYL: Needles from local pine trees 10 times heavier than normal; SIBERIA: High mortality and cancer rate found in people near former bomb test site; people's radioactivity as high as those near Chernobyl.

(6) SWALLOW INVASION: ITALY; Birds flocked into area to feast on gnats attracted to algae infesting lagoons; air traffic stalled.

(7) FREED BIRDS: CHINA: One man has freed 3,000 caged birds he bought over five years; INDIA: 4,000 birds freed at Gandhi's grave to urge people not to buy birds.

(8) EARTHQUAKES: SOUTHERN MEXICO, 5.5 mag.; PHILIPPINES, 5.9 mag.

(9) Aug. 17: FULL MOON; Total eclipse of the moon.

Solar Geomagnetic Data

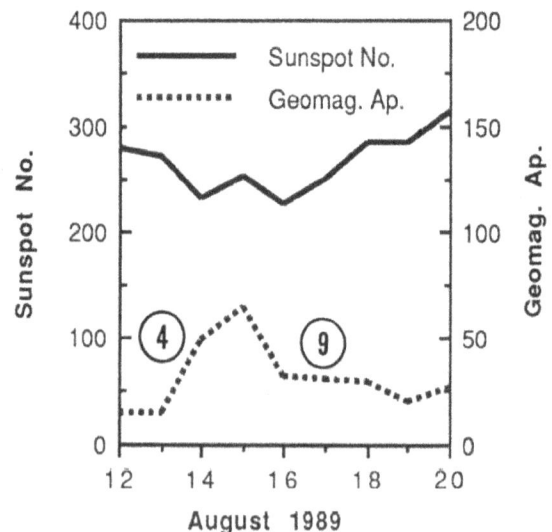

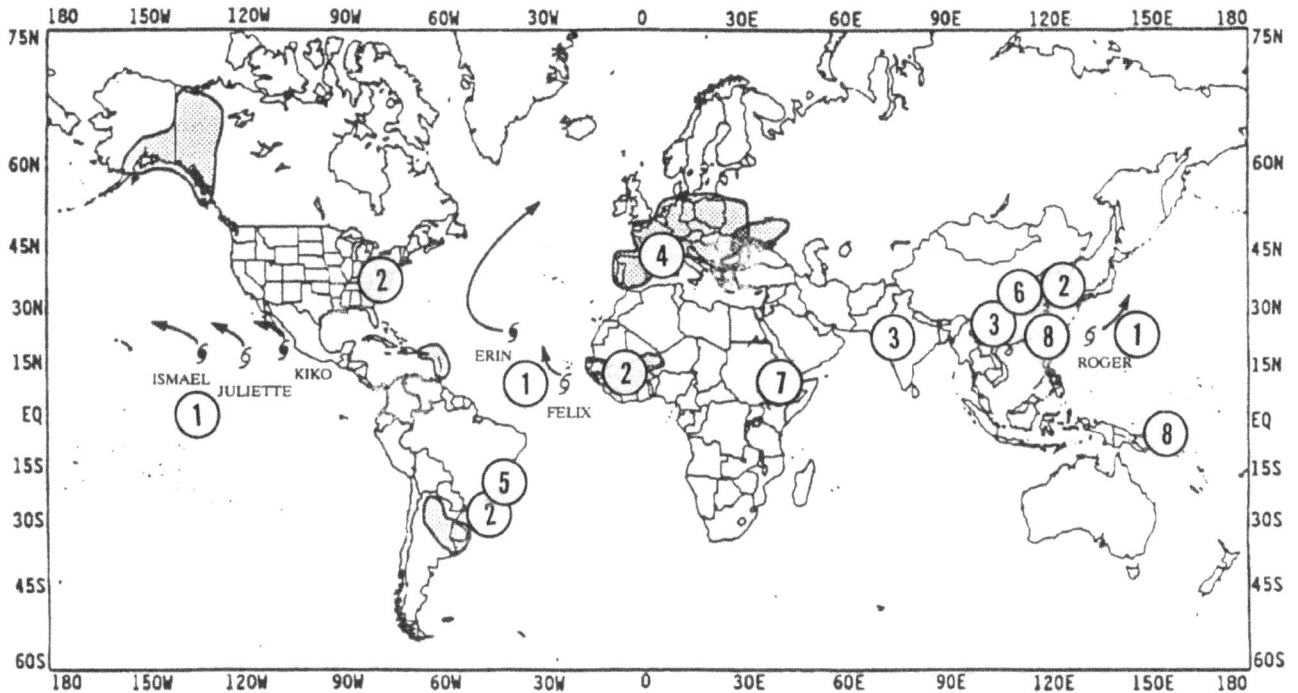

FOR THE WEEK OF AUGUST 26, 1989

Persistent Conditions (shaded)

W. CANADA, ALASKA: Warm weather retreats in some areas (4 weeks).

CARIBBEAN ISLANDS: Dry conditions persist (10 weeks).

N. ARGENTINA, URUGUAY, S. BRAZIL: Unusually warm temperatures subside (ended at 2 weeks).

EUROPE: Heat wave continues; hot, dry winds from the Sahara blow into the region (6 weeks).

ROMANIA, MOLDAVIAN, UKRAINIAN SSR: Dry conditions continue (8 weeks).

Transient Events (numbered)

(1) TROPICAL STORMS: Three storms in eastern Pacific, *Kiko* causes damage over Baja; storm in west skirts Japan; two storms in Atlantic head out tosea.

(2) HEAVY RAINS: S. KOREA: 6 dead, 1,200 homeless from floods; E. USA. Rains brought many rivers to flood level; third wettest year since records kept in New York: W. AFRICA: Torrential rains cause flooding; BRAZIL: Buenos Aires hit by floods, 10,000 homeless.

(3) DROUGHTS: CHINA: Areas report crop losses, water shortages; INDIA: Western part of country in drought, bird sanctuaries endangered.

(4) HEAT WAVE: FRANCE: Grape harvesting early due to hot, dry weather.

(5) OIL SPILL: BRAZIL: 33,800 gallons spilled onto beach when pipeline punctured; 7,500 gallons dumped into ocean by leaky tanker.

(6) FLY KILLING: CHINA: Man kills 4,000 flies in Beijing daily; government buys them to encourage killing.

(7) Aug. 20-21: EARTHQUAKES; ETHIOPIA, 6.3 mag., 2 people killed, damage reported.

(8) Aug. 21: EARTHQUAKES; SOLOMAN ISLANDS, 6.0 mag.; TAIWAN, 6.2 mag.

Solar Geomagnetic Data

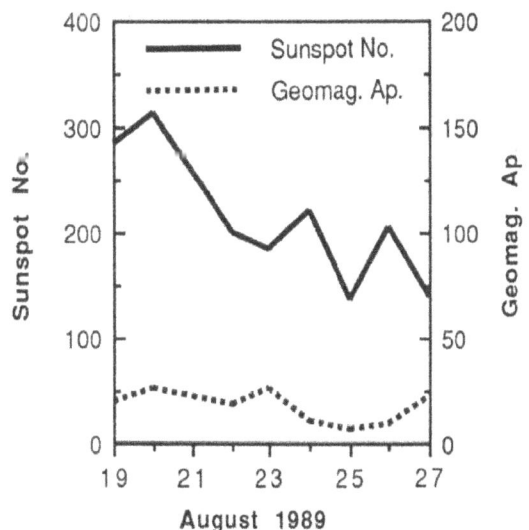

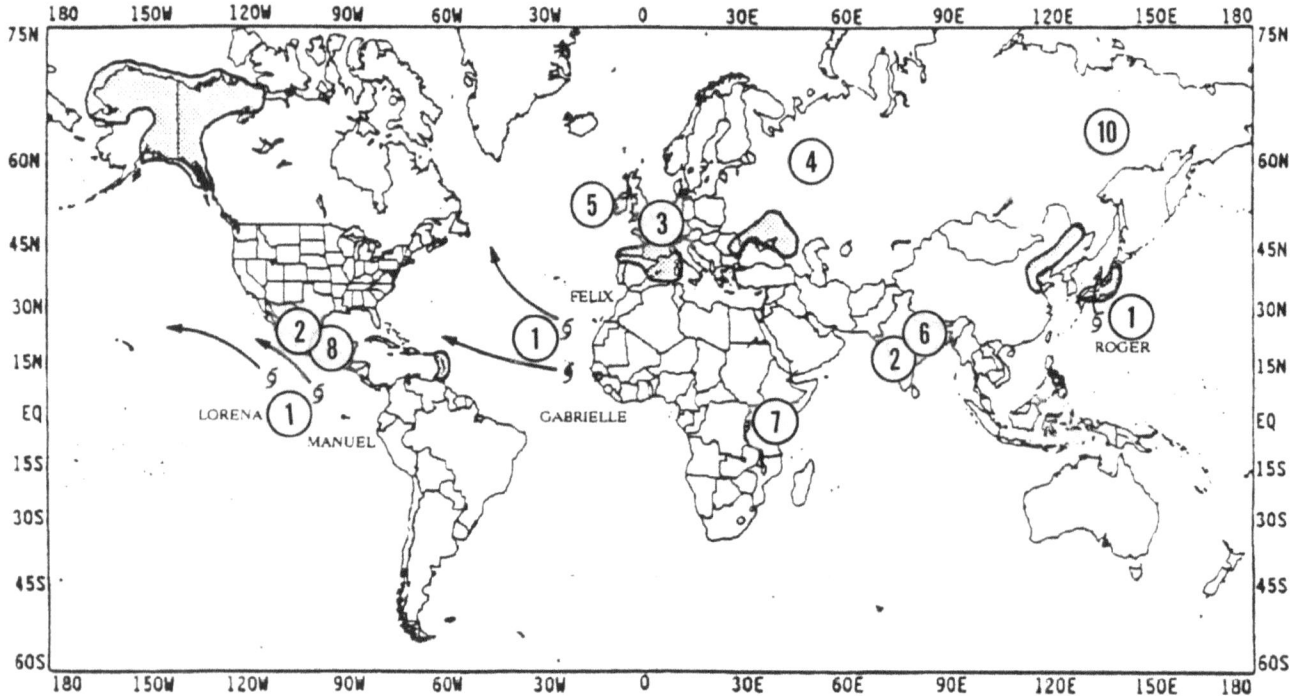

FOR THE WEEK ENDING SEPTEMBER 2, 1989

Persistent Conditions (shaded)

W. CANADA, ALASKA: Unusually warm weather continues (5 weeks).

CARIBBEAN ISLANDS: Rains bring limited relief from dryness (11 weeks).

EUROPE: Cool, wet weathers diminishes heat (ending at 7 weeks).

BULGARIA, ROMANIA, MOLDAVIAN, UKRAINIAN SSR: Rains bring some relief, but dryness continues in most areas (9 weeks).

NE CHINA: Drought conditions persist (5 weeks).

Transient Events (numbered)

(1) TROPICAL STORMS: *Roger* hits Japan with more rain; two storms form in eastern Pacific and head to sea; two storms form off Africa and head west.

(2) HEAVY RAINS: INDIA: Mudslides and flooding: MEXICO CITY: 60,000 people homeless north of city due to floods.

(3) LIGHTNING: FRANCE: 12 cows, five calves, and bull killed while under tree.

(4) NUCLEAR CONTAMINATION: USSR: Lake in Ural mountains contaminated by waste dumping with 2.5 times amount of radiation released from Chernobyl.

(5) RAT PLAGUE: BRITAIN: 20 % rise in rat population due to mild weather.

(6) BIRD DEATHS: INDIA: Birds appearing to commit suicide in jungles may actually be disoriented by either fog or region's geomagnetic fields.

(7) DRUNK PIGS: KENYA: After drinking local brew, 13 pigs ran amok in town, chasing children and damaging homes.

(8) Aug. 19: EARTHQUAKES; MEXICO, 6.6 mag.

(9) Sept. 2: ATOMIC BOMB; USSR, Semipalatinsk, magnitude not available.

Solar Geomagnetic Data

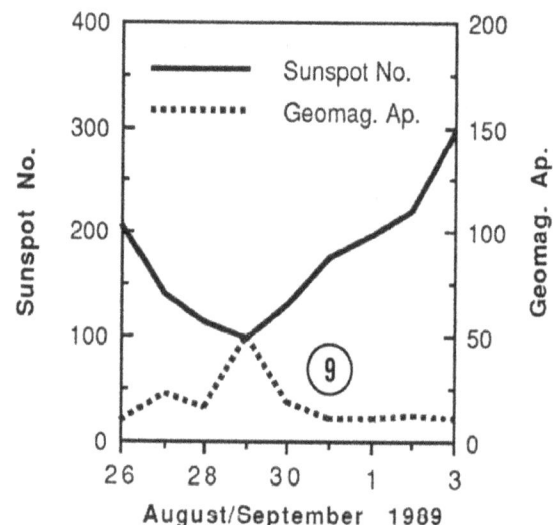

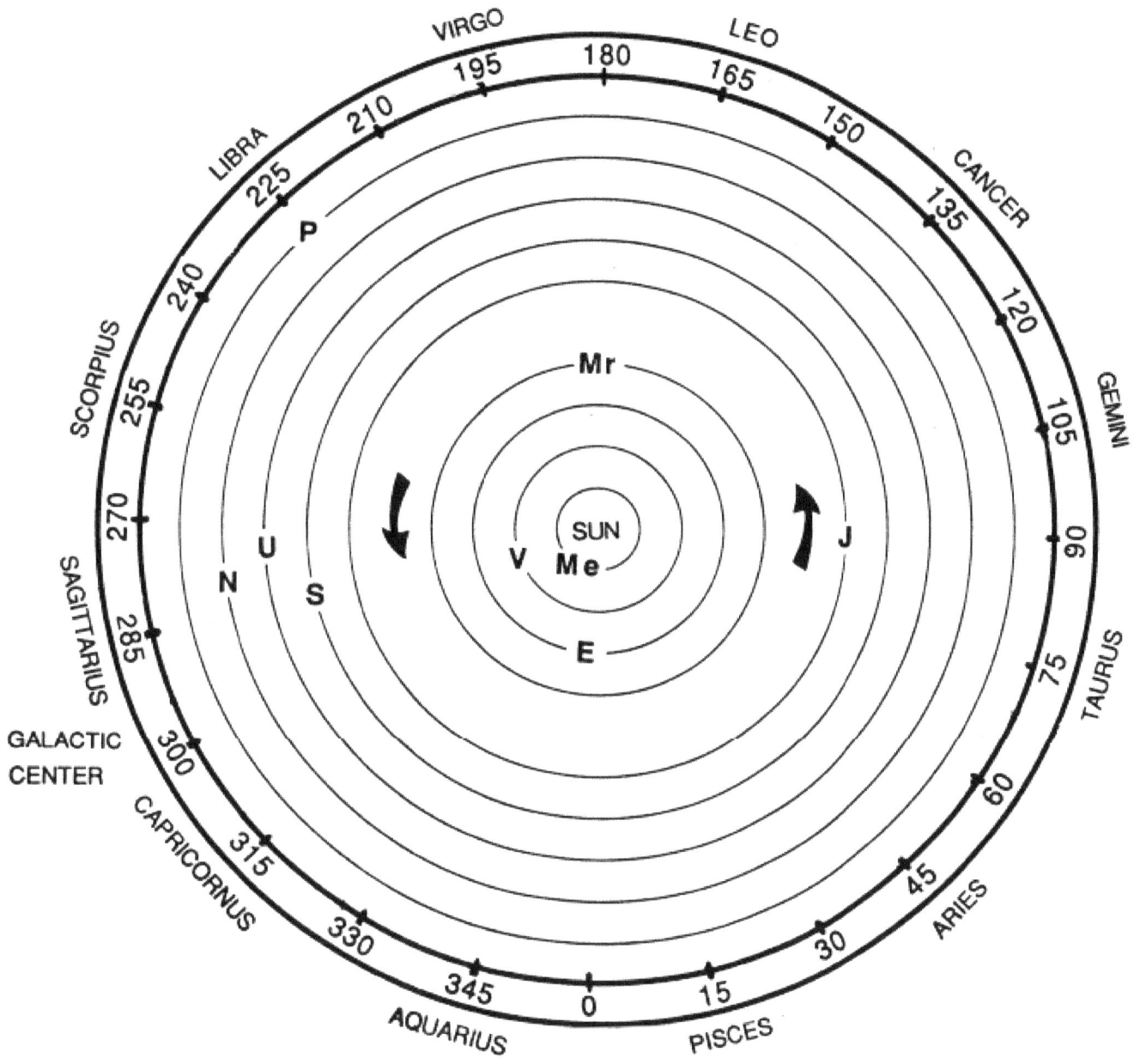

Notes: 1) Planetary distances from the Sun are not in correct proportions, especially for the outer planets. Angles between planets are therefore only approximate.

2) Planets move in a counter-clockwise direction, spiraling out of the page, towards the observer, as the Solar System moves through the heavens.

3) True distances from the Sun, in astronomical units, and approximate revolutionary period.

Mercury	Me	0.38 AU	88	days to orbit once, or 3.7 days per 15 degrees
Venus	V	0.72 AU	225	days to orbit once, or 9.4 days per 15 degrees
Earth	E	1.00 AU	365	days to orbit once, or 15.2 days per 15 degrees
Mars	Mr	1.5 AU	687	days to orbit once, or 28.6 days per 15 degrees
Jupiter	J	5.2 AU	11.9	years to orbit once, or 181 days per 15 degrees
Saturn	S	9.5 AU	29.5	years to orbit once, or 1.2 years per 15 degrees
Uranus	U	19 AU	84.0	years to orbit once, or 3.5 years per 15 degrees
Neptune	N	30 AU	164.8	years to orbit once, or 6.9 years per 15 degrees
Pluto	P	39 AU	247.7	years to orbit once, or 10.3 years per 15 degrees

✳ The May 1988 issue of the *Journal of Orgonomy* carried the first part of a major article by James DeMeo, on "Desert Expansion and Drought", which suggested that the large desert region of *Saharasia* (North Africa, Middle East, Central Asia) is a source region for the development and maintenance of desert regions at a great distance. This thesis was suggested when areas of desert were mapped along with areas of seasonal drought, or "dry periods". Given that desert lands are enlarging at around 70,000 square kilometers every year, seasonal droughts are becoming more lengthy and intense. A map constructed by DeMeo, and published in the article, demonstrated the interconnected nature of all the world's desert regions. "The map gives the chilling impression of a planet being attacked by a large, growing cancer tumor. The major expression of this 'planetary cancer' is across the same hyperarid territory I previously have identified as Saharasia, which did not exist prior to c.4000 BC. Secondary, connected desert regions also exist, similar to metastases." (See Figure 1 below.)

✳ An unusual aspect of droughts that affected North America within recent years has been identified by workers with the National Climate Data Center. A graph showing the monthly percent of the USA experiencing severe drought shows the following features: 1) A regular summertime increase in drought, in a cyclical manner and; 2) an additional drought feature that began in the summer of 1987. Indeed, the drought conditions of 1988 appear to have started during the summer of 1987, and peaked out in 1988. The small arrows in Figure 2 indicate the various cycles.

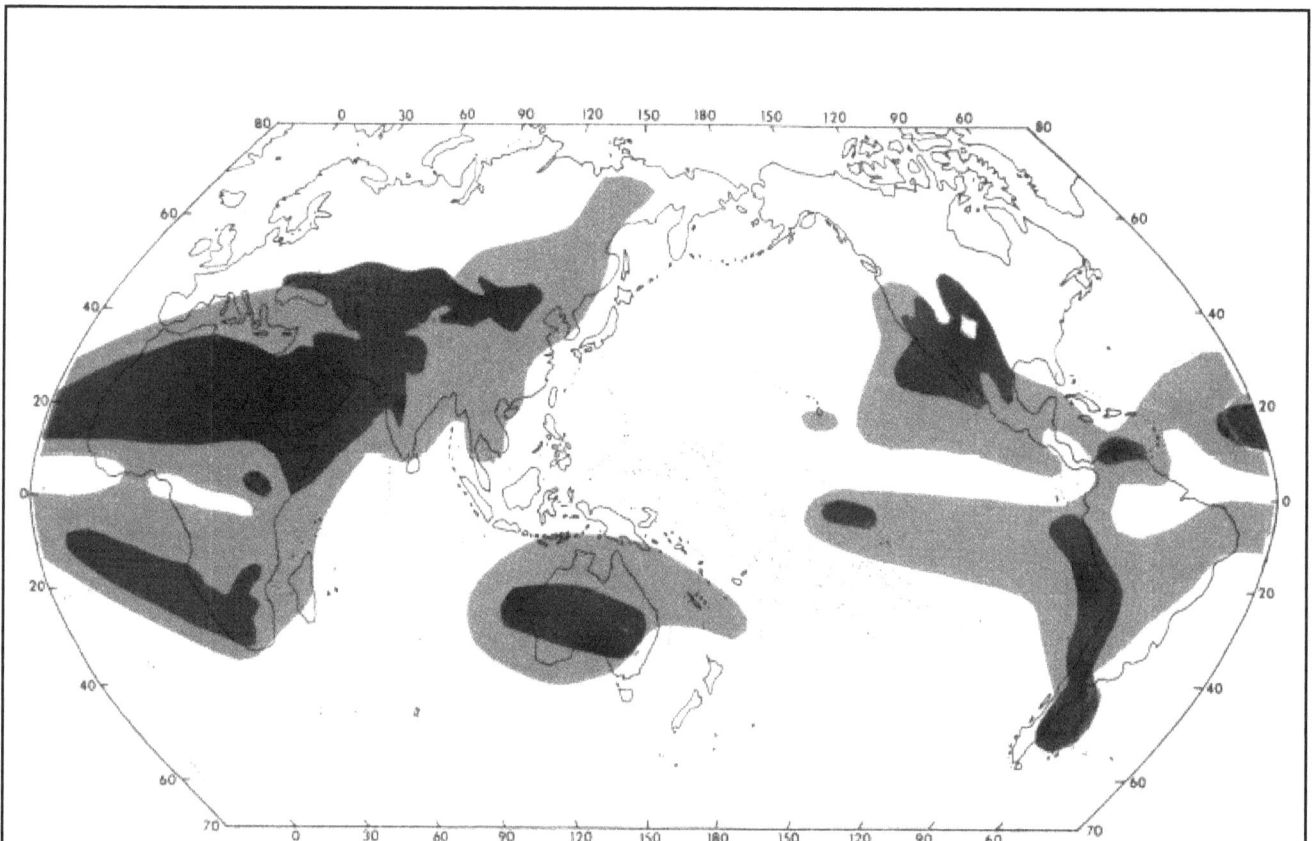

Figure 1. DESERT/DROUGHT MAP. Desert regions over land and ocean appear in black (Koppen's BW or BS climate types). Shaded or grey areas mark regions that experience a regular dry season (Koppen's Aw, Cs, Cw, and Dw climate types).

U.S. % AREA IN SEVERE / EXTENDED *DROUGHT*
January 1984 through June 1989

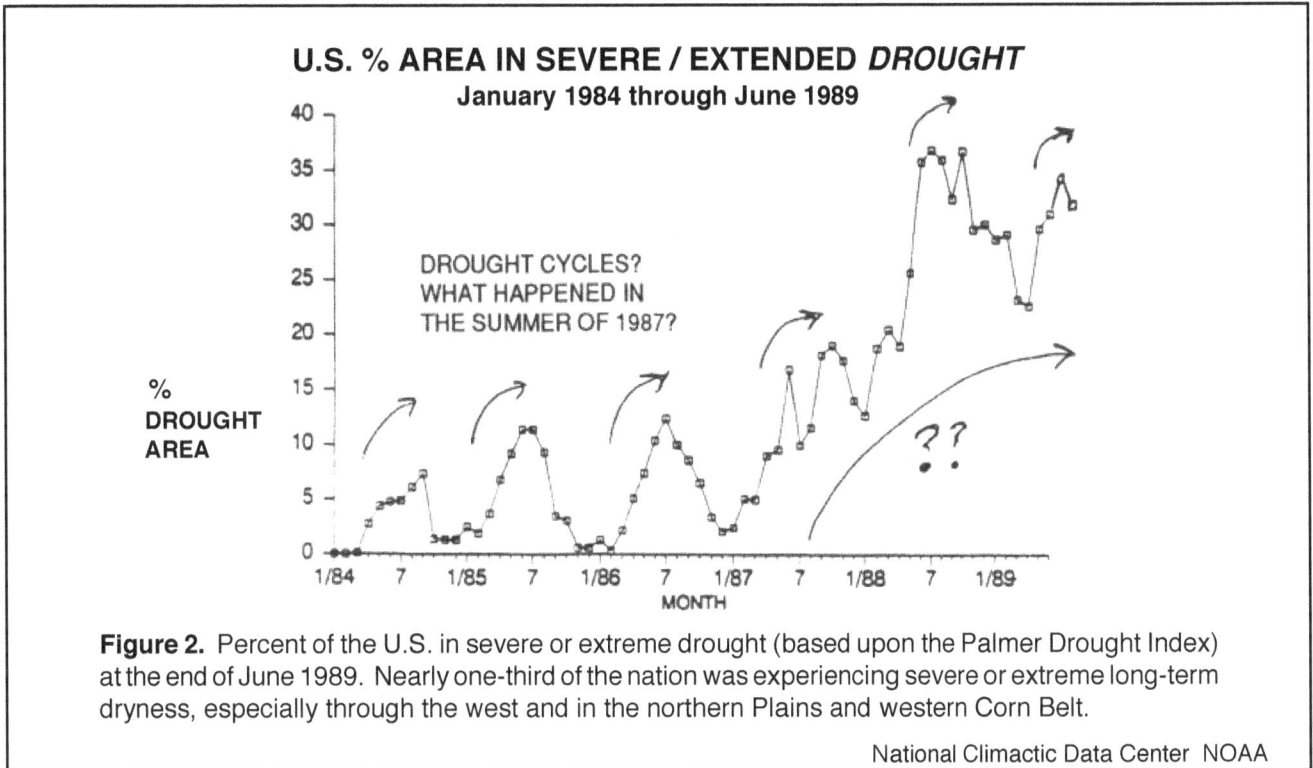

Figure 2. Percent of the U.S. in severe or extreme drought (based upon the Palmer Drought Index) at the end of June 1989. Nearly one-third of the nation was experiencing severe or extreme long-term dryness, especially through the west and in the northern Plains and western Corn Belt.

National Climactic Data Center NOAA

∗ The new edition of *The Orgone Accumulator Handbook: Construction Plans, Experimental Use, and Protection Against Toxic Energy*, by James DeMeo, Ph.D., will be available for sale in late October. The book contains a foreword by Eva Reich, M.D., and is over 150 pages, paperbound, with a full color cover showing the vividly-glowing blue orgone energy field of an Apollo astronaut walking on the lunar surface. (Natural Energy Works, Ashland, Oregon, USA. www.naturalenergyworks.net)

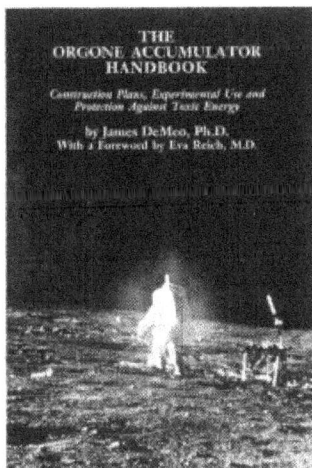

∗ Weekend workshops on "The Bioenergetic, Orgonomic Basis of Life and Weather" were held in Cambridge, Seattle, and Berkeley in 1989, but scheduled workshops in Atlanta and Boulder were cancelled, due to lack of enrollment. Myron Sharaf and Renate Reich Moise (Reich's Granddaughter) participated at the Cambridge workshop, and enlivened that event considerably by providing many personal memories and anecdotes about Wilhelm Reich. Grier Sellers, who is currently studying clinical acupuncture, assisted with a demonstration on the formation of sand bions, and on the bionous disintegration of blood, at the Berkeley workshop. Our thanks to all who helped make these events so interesting.

∗ Funding for necessary Laboratory projects continues to be a major obstacle. We have not yet found a sponsor for the Desert Greening Project, but are hopeful that the evidence gathered from this year's field work will open the door to such funding. Funds are also needed for satellite imaging equipment, additional computer equipment, for a binocular microscope to initiate necessary bion studies, for video recording equipment to document our field operations, for additional miscellaneous laboratory and office equipment, research data, and basic operating expenses. A listing of these items is given on page 93 of the *Pulse*. Call or write for details if you are in a position to help out.

✳   Last year, in August of 1988, a cloudbusting field expedition was undertaken in the desert regions of the American Southwest, between Los Angeles and Phoenix. This operation had two goals: 1) to break up the desert dor layer and bring increased clouds and rains to the desert, and 2) to restore the galactic orgone energy stream ("subtropical jet") into the Great Plains and Midwest, and thereby also restore rains to that droughty area. We are happy to report that both goals were achieved. In September of 1988, a second cloudbusting operation was launched into the drought regions of the State of Washington. That operation restored rainfall to the State, and in conjunction with other cloudbusting efforts by workers in Idaho, stimulated rainfall very far inland. The raging fires in Yellowstone National Park were extinguished around this time, and the drought which had affected the Pacific Northwest for at least a year was finally broken. The figures below summarize the resulting cloud cover and rainfall data in Arizona and Washington States, obtained from the National Weather Service. A full report of these operations appeared in the May 1989 issue of the *Journal of Orgonomy* (V.23,#1), under the title "Breaking the Drought Barriers in the SW and NW United States".

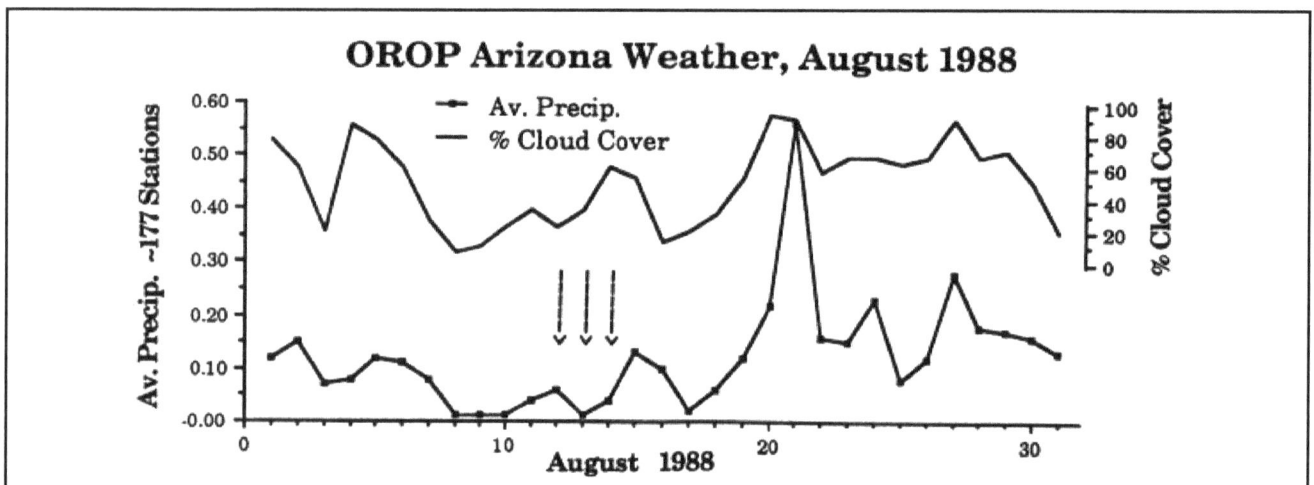

Figure 1: Precipitation and Percent Cloud Cover, Arizona, August 1988. National Weather Service data. Cloudbusting operations ran from August 12-14, as noted by the arrows. Note the periods of high percent cloud cover, but little rain, in the early part of the month, but the large peak in rainfall that occurs shortly after the cloudbusting operations. That peak in rainfall coincided with the first major rains and cool temperatures to occur in the Midwestern USA in many months, and signaled a major break from drought conditions.

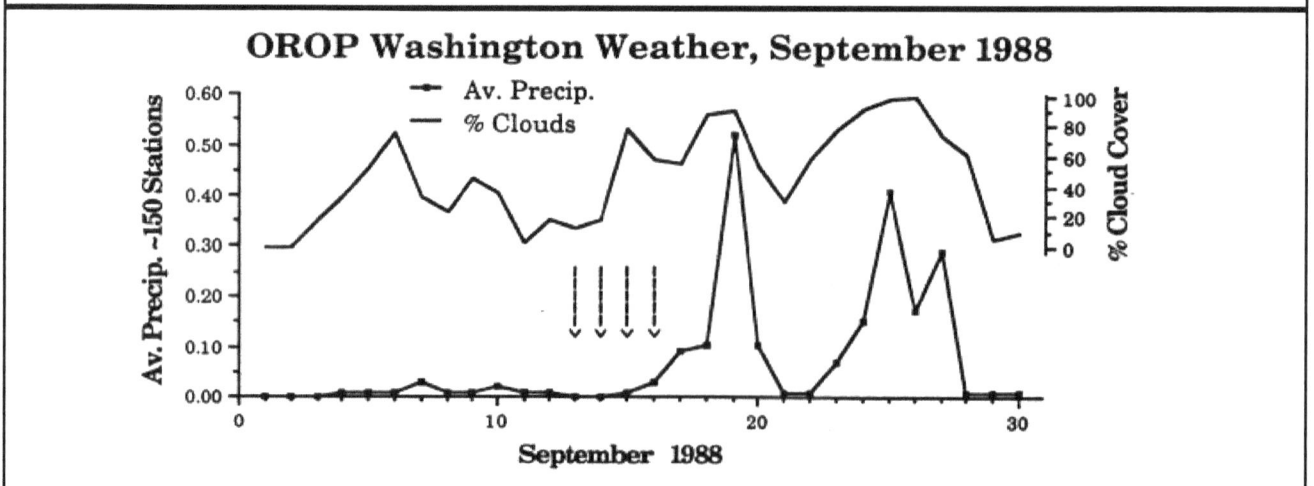

Figure 2: Precipitation and Percent Cloud Cover, Washington, September 1988. National Weather Service data. Cloudbusting operations ran from September 13-16, as noted by the arrows. Note the periods of high percent cloud cover, but no rains, in the early part of the month. The figure clearly shows the restoration of atmospheric pulsation following cloudbusting operations.

* A series of cloudbusting operations were undertaken in the San Francisco Bay area early in 1989, at a station maintained by the American College of Orgonomy. These operations were followed by a resurgence of moisture into northern California, and greatly increased the snowpack across the northern Sierras. Restrictions on water usage in the San Francisco Bay area were eliminated in subsequent months, as reservoirs filled. See the CORE Reports section of the November 1989 *Journal of Orgonomy* for details.

* Five separate cloudbusting field expeditions were made into the dryland regions of the American Southwest this summer (1989), between Los Angeles and Phoenix. While those experiments await a more complete analysis, using National Weather Service data, preliminary indications are that the results were both widespread and significant. Parts of Arizona close to the operating cloudbuster received up to 500% of normal rains, and each of the cloudbusting operations was followed by a significant pulse of energy and moisture into the Great Plains and Midwest. Indeed, the threat of forest fire and drought greatly diminished across the USA in 1989, as compared to the previous two years. On a few occasions, these operations were assisted by secondary draws by a second cloudbuster located in the San Francisco Bay area, and by a third in Kansas. A preliminary report on these experiments will appear in the November issue of the *Journal of Orgonomy,* with a more in-depth report in the May 1990 issue. Our thanks to the American College of Orgonomy, which provided the major portions of the funding for these operations, and to Mr. Douglas Harper and a friend overseas for additional funds. Our thanks also to the many people who assisted this work in both large and small ways, particularly to those who came along and roasted in the dry desert heat with us. Additional information on these important desert experiments will also appear in subsequent issues of the *Pulse.* (See Figure 4 on the following page.)

U.S. % AREA UNUSUALLY MOIST
JANUARY 1984 THROUGH JUNE 1989

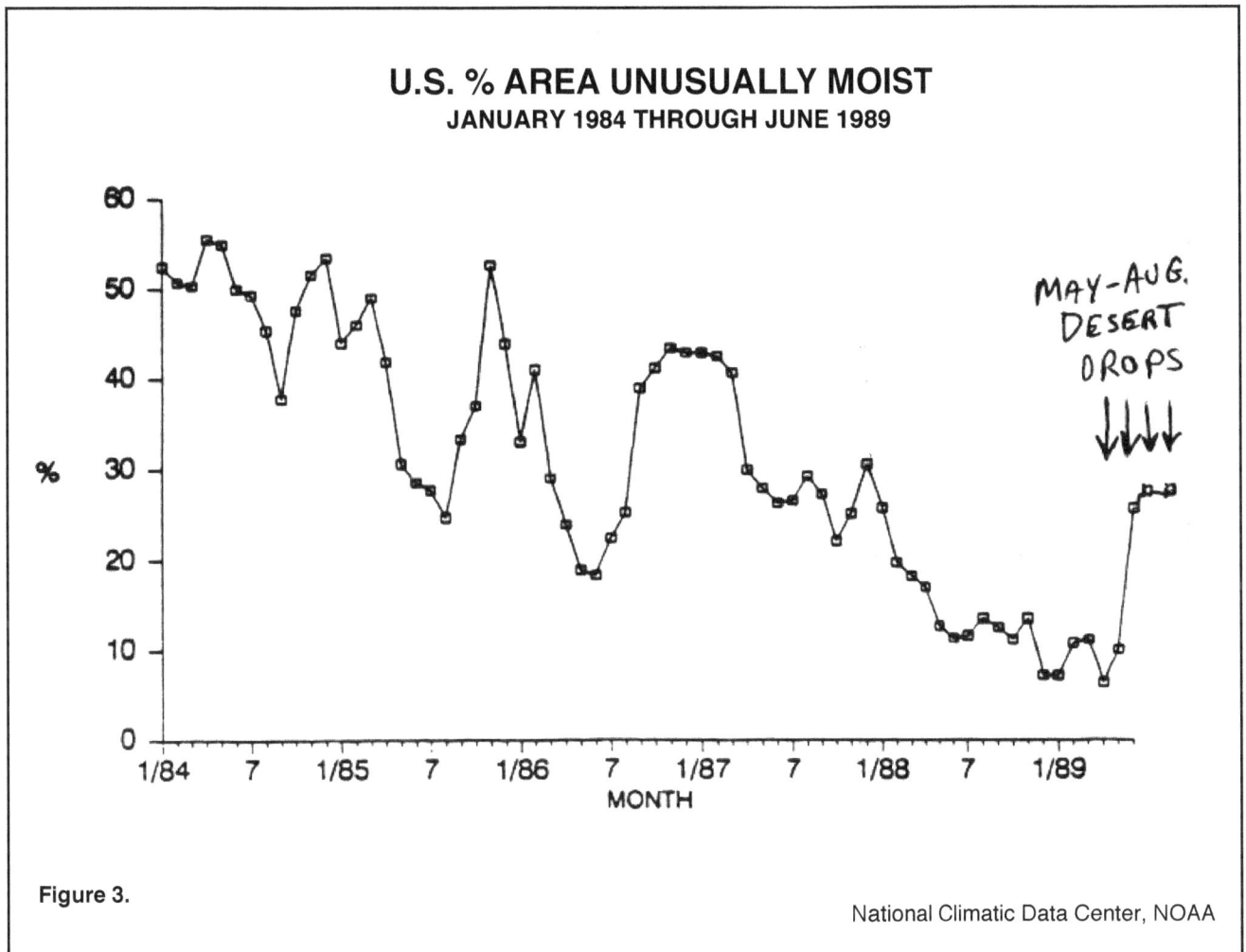

Figure 3.

National Climatic Data Center, NOAA

PERCENT OF NORMAL PRECIPITATION

Summer
(June - August 1989)

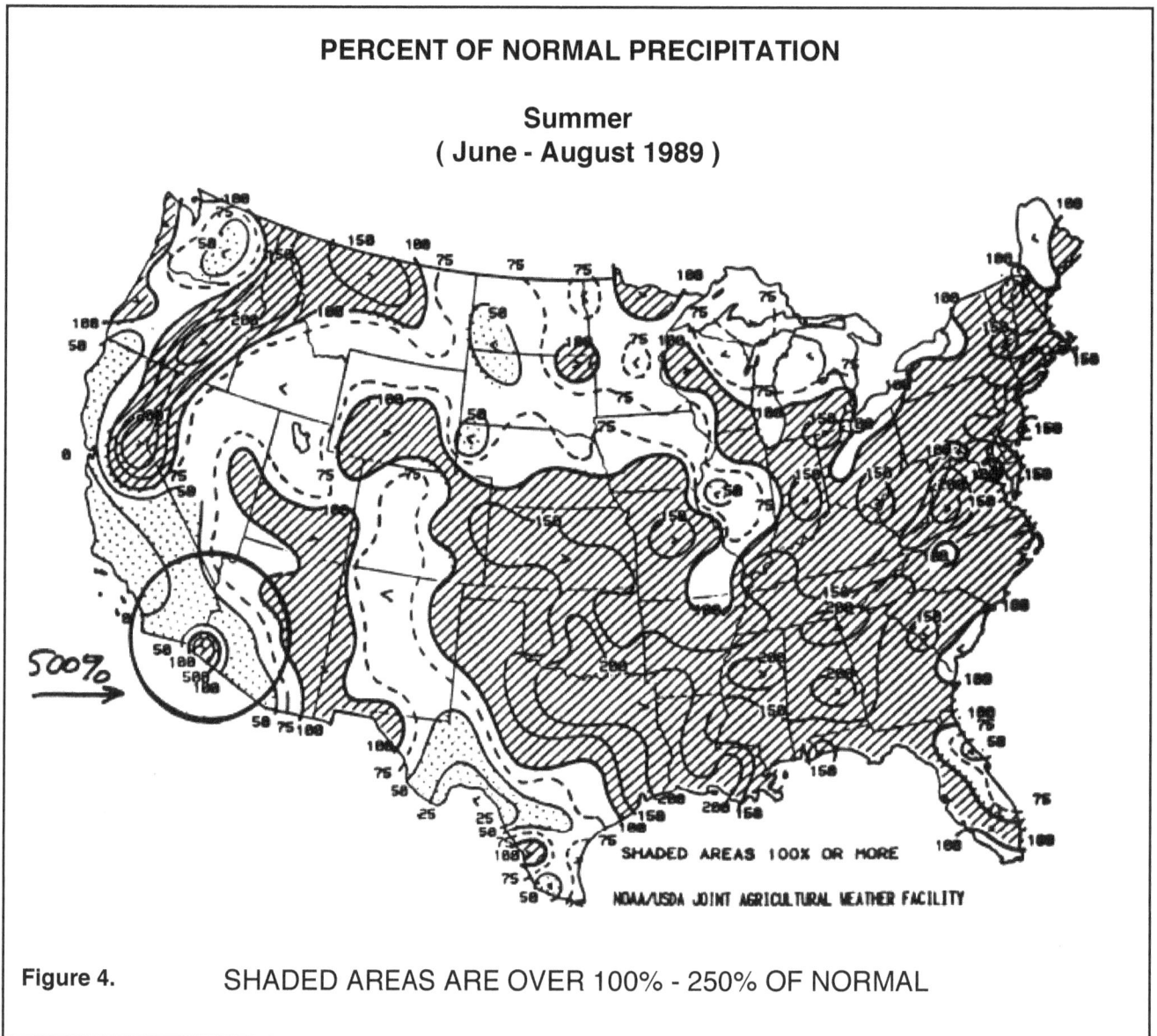

SHADED AREAS 100% OR MORE

NOAA/USDA JOINT AGRICULTURAL WEATHER FACILITY

Figure 4.   SHADED AREAS ARE OVER 100% - 250% OF NORMAL

✶ The Orgone Biophysical Research Laboratory may undertake an experimental cloudbusting program in West Germany, to address the problem of forest death. There is some evidence to suggest that forest death may primarily be a problem of atmospheric stagnation and dor energy, and that the cloudbuster can affect a beneficial change in the atmosphere where trees are dying. This includes a possible beneficial change in the pH of rains that follow cloudbusting operations. This possibility was previously discussed in a paper by DeMeo, in the *Journal of Orgonomy* ("Reduction of Rainwater Acidity Following the End of the 1986 Drought: An Effect of Cloudbusting?").

1989
∗ **December 27:** Princeton, New Jersey, Directed Reading on Friedrich Nietzsche, Dr. Charles Konia, contact the American College of Orgonomy, PO Box 490, Princeton, NJ 08542, telephone (201) 821-1144.

1990
∗ **January - May**: Cambridge, Massachusetts, Seminar on "The Integration of Analytic and Reichian Approaches in Psychotherapy", led by Myron Sharaf, Ph.D.; For more information, contact Dr. Sharaf at (618) 969-1653.

∗ **April 7-8**: Berkeley, California, Introductory Weekend Workshop on "The Bioenergetic, Orgonomic Basis of Life and Weather", led by James DeMeo, Ph.D.; For more information, contact Orgone Biophysical Research Lab, PO Box 1395, El Cerrito, CA 94530, telephone (415) 526-5978.

∗ **April 21 - 24**: Princeton, New Jersey, Advanced Laboratory Workshop on Orgone Biophysics is offered by the College of Orgonomy. For more information contact the ACO, PO Box 490, Princeton, NJ 08542, telephone (201) 821-1144.

∗ **April 22:** Planet Earth, Earth Day Global Activities; For information on local events in your area, contact your local environmental groups, or write Earth Day 1990, PO Box 96773, Washington, DC 20077 USA.

∗ **June 15-18:** Nice, France, Fifth International Orgonomic Conference, Palais des Congres - Acropolis, various speakers and topics, Contact Dr. Giuseppe Cammarella, SEDIFOR, Allee due Chene Vert, Parc Liserb, 06000 Nice, Cimiez - France, telephone 93.81.96.96, or the American College of Orgonomy, PO Box 490, Princeton, NJ 08542, USA, telephone (201) 821-1144.

∗ **July**: Los Angeles, California; Annual Cancer Control Society Convention, various speakers and subjects, Date not firmly set at the time of this publication, Contact Lorraine Rosenthal, 2043 N. Berendo St., Los Angeles, CA 90027, telephone (213) 633-7801.

∗ **August 26-September 3**: Vienna, Austria, 12th International Congress of Biometeorology, Study group on Physiochemical-Biological Fluctuating Phenomena (Piccardi Group); Contact Dr. E. Wedler, Meteorologie, Frie Univ. Berlin, Dietrich Schafter Weg 6-10, D-1000 Berlin 41.

∗ **October**: Bremmen, West Germany, Conference on Wilhelm Reich, University of Bremen, various speakers and subjects. Dates not set at the time of this publication. Contact Dr. Heiko Lassek, Delbruchstr. 4-C, 1000 Berlin 33, West Germany, telephone 030-891 49 14.

∗ **October 20-21**: Berkeley, California, Introductory Weekend Workshop on "The Bioenergetic, Orgonomic Basis of Life and Weather", led by James DeMeo, Ph.D.; For more information, contact Orgone Biophysical Research Lab, PO Box 1395, El Cerrito, CA 94530, telephone (415) 526-5978.

NOTE: Workshops or lectures on various aspects of orgonomy can be made at other locations by special arrangement. Contact the Orgone Biophysical Research Laboratory, or the American College of Orgonomy for details (addresses and telephones given above).

Appearing in the *Journal of Orgonomy* 23(1), May 1989:
(PO Box 490, Princeton, NJ 08542)

- Further Problems of Work Democracy (IV), by Wilhelm Reich, M.D.
- In Seminar with Dr. Elsworth Baker
- Desert Expansion and Drought: Environmental Crisis (Part I), by J. DeMeo, Ph.D.
- Work Energy and the Character of Organizations (Part I), by Martin Goldberg, M.S.
- The Creation of Matter in Galaxies (Part II), by Charles Konia, M.D.
- Finger Temperature Effects of the Orgone Accumulator, by N.Snyder, M.S.W., Ph.D.
- Orgone Therapy (VIII: Functional Thinking in Medical Practice), by C. Konia, M.D.
- Music and Emotional Expression in Whitman's *Leaves of Grass*, by J. Yordy, B.S.
- How to Integrate an Unknown Function, by Jacob Meyerowitz, B. Arch.
- CORE Report #19: Eastern US Weather Operations (6/1/88 - 7/21/88),
  by John Schleining, M.S.
- CORE Report #20: Breaking the Drought Barriers in the SW and NW USA,
  by James DeMeo, Ph.D.
- Response to Martin Gardner's Attack on Wilhelm Reich and Orgone Research in
  the *Skeptical Inquirer*, by James DeMeo, Ph.D.

Appearing the *Pulse of the Planet* No.1, Spring 1989:
(PO Box 1395, El Cerrito, CA 94530)

- Cloudbusting, A New Approach to Drought, by James DeMeo, Ph.D.
- Recent Abnormal Phenomena on Earth and Atomic Power Tests, by Yoshio Kato
- Response to Martin Gardner's Attack on Wilhelm Reich and Orgone Research in
  the *Skeptical Inquirer,* by James DeMeo, Ph.D.
- Postscript on the Food and Drug Administration's Evidence Against Wilhelm Reich,
  by James DeMeo, Ph.D.

Appearing in Other Publications:

- "Big Brother: FDA, Excerpts from the *Congressional Record*, 1965", *Newsletter*,
  Friends of the Wilhelm Reich Museum, #25, Spring 1989, Rangeley, ME.
- "The Geography of Genital Mutilations", by James DeMeo, Ph.D., *Truth Seeker*,
  July/August 1989, p.9-13.
- "The Doctor Who Made it Rain", by Tim Clark, *Yankee Magazine*, September 1989,
  pp.72-79,130.

NOTE: The Wilhelm Reich Museum will soon begin publication of a new periodical titled *Orgonomic Functionalism*, containing many previously out-of-print articles by Dr. Wilhelm Reich. For details, contact the Wilhelm Reich Infant Trust Fund, 382 Burns St., Forest Hills, NY 11375.

ORGONOMIC OBSERVATIONS

Pin Points in the Sky?

Editor's Comment: *A recent letter to the Pulse about the visible orgone units rekindled interest in published reports of the observable phenomena outside of journals devoted to orgonomy. Have people outside of orgonomy written about this visible expression of the atmospheric orgone energy? The answer to this question is yes! The following series of "Letters to the Editor" appeared in the New Scientist magazine, in September of 1957. They addressed the question of the visualization of the pointed form of the atmospheric orgone energy, what Wilhelm Reich called the orgone units. The correspondence came to our attention after being previously published some years ago in England, in the November 1957 issue (IV:6) of Orgonomic Functionalism, edited by Paul Ritter (not to be confused with the new journal of the same title planned for publication by the Wilhelm Reich Museum). The discussion appropriately points to the possibility that some of the visual phenomena may indeed be an affect of blood coursing through vessels in the eye. This explanation is made quite clear when such visual luminescent phenomena move in a manner timed to one's own heartbeat. However, other forms of this phenomena are not so easily explained, appearing much like luminous champaign bubbles, or like fogs, flames or waves, the movement of which may be dependant upon environmental factors (weather, local energetics). When these visible phenomena can additionally be magnified, as demonstrated by Reich with the orgonoscope, their objective nature "outside the eye" is clarified. The correspondence below additionally shows how far some are willing to go to "explain away" this observable phenomena without attempting to investigate it. We may contrast this to Reich's approach, which was to study the "subjective light phenomena" at length, leading to the development of new techniques and instruments for objectifying it.*

Sir,

I have noticed that when one looks up at a clear patch of sky on a sunny day, focusing in mid air as it were, several interesting visual phenomena occur. 1) Countless bright "pin points" can be seen whirling around in apparently complicated dance patterns. 2) Small "discs" with a black dot nucleus and ring-like periphery, move slowly across the field of vision. 3) Faintly grayish "wisps" may be observed from time to time, moving in a downward and sideways direction. Points 2) and 3) obviously have some relationship with the eye itself, as any movement of the eye also affects the movement of these two phenom-

ena. The bright "pin-points" however, would seem to be objective, as they are quite uninfluenced by any eye movements. I wonder if any of your readers can satisfactorily explain just what is taking place?
- J. H. Chadwick, Manchester 5 September 1957

Sir,

Your correspondent, J. H. Chadwick (Letters, 5 September), is observing, not flying saucers, but corpusculae and muscae volitantes. The first are his "pin points" which follow the course of the blood vessels in the retina of the eye. Being anterior to the light-sensitive nerve layer, a shadow is cast which is seen and projected into space. The origin may be blood corpuscles or possibly pulse pressure waves stimulating the nerve endings directly. The movement is sinuous towards and away from the point of regard in space which corresponds to the macular region of the retina. There are no blood vessels in the macular area, and it will be observed that the "pin points" disappear as they approach the point of regard. Mr. Chadwick's "discs" with a black nucleus and his greyish "wisps" may be caused by floating cells detached from the membranes of the eye, suspended in the vitreous in the posterior chamber, in the fluid in the anterior chamber, or commonly in the tears on the surface of the eye. Unless numerous and quite efficient they are of no significance. If the obstruction is transparent but of a different refractive index from the surrounding medium, refraction might cause an appearance of a dark nucleus and a light halo or a dark ring with a light centre.
- W. Kenneth Beard, Cornwall 12 September 1957

Sir,

I read with great interest the letter of Mr. J. H. Chadwick headed "Pin points" in the sky (5 September), as I have observed the same phenomena as your readers, although I had assumed it to be a peculiarity of my vision, as mere hallucination. It would be, I think, possible that particles suspended in the fluid secreted by the lachrymal gland are received by the retina through a complicated system of refection. This system must produce a large magnification as the particles viewed are of a very small order. The "discs" referred to could then be explained by saying that they are in actual fact circular particles, the outer edges not being focused correctly on the retina, thus causing an inner defined dot with an outer undefined ring as penumbra. Again the greyish (sic) "whips" are hair-like particles treated in the same manner, thus causing the rather ghostly appearance which they give. The theory that they are particles in the first place is strengthened by the fact that they move in relation to the eye-ball. When the eye is moved upwards the particles disappear, and when the eye is at rest they slide downward, showing that they are suspended in the lubricating fluid of the eye. The effect of the pin points is very odd, and the only explanation that I could offer would be that minute air bubbles are continuously breaking up in the fluid. This, I admit, is not very feasible. The reason why your reader sees these things when looking up at the sky is that there are no dark objects to fall on the retina, and so distract attention from the other happenings. Again, focusing the eye at the considerable distance may have a lot to do with it. - G. H. Watkins, Middlesex, 12 September 1957

Sir,

Mr. Chadwick has noticed "phenomena" that I did not notice until about two years ago (I am 77). If he has seen the blood circulating in a frog's foot under a microscope as I did at school, he can realize as I do that the "whirling pin points" are the result of a microscopic action in the eye working on a thin transparent membrane. You are seeing a portion of your own blood circulating and the corpuscles therein.

- F. Newell, London, 12 September 1957

Sir,

The matter of the "pin points" of light with Mr. J. H. Chadwick raised (5 September) is not as simply explained as the three correspondents who answer would have it. Although it may be a shock to them the pin-points have been magnified both by myself and others and are therefore, as Mr. Chadwick suggests, "objective" and external to the eye.   Of these I have not been able to find any explanation in the orthodox physics or physiology.

- P. Ritter, Nottingham 26 September 1957

Sir,

Your correspondents interested in the explanations of the "pin points" in the sky may be amused to hear of a curious book devoted to the same problem which appeared many years ago. It was called *Etheric Vision*, and its author, H. Desmond Thorp, passed the time while a prisoner of the Kaiser experimenting — if that is the word — with these specks of light, which he assumed had objective existence. He named them "ionites", and claimed that they were the foci of some form of energy in space. Thorp dismissed the notion that his ionites were a product of the life of the eye because among other things, they seemed to move in three dimensions. After much practise, he tells us, he succeeded by an effort of concentration to bring one such point to a standstill. The surprising effect of this was that it began to expand, becoming so inflated that it finally burst! He repeated the experiments many times, with the same results. He also discovered that, by fixing his attention on a single ionite, he could will it to move in any direction he wished. The book impressed

me as being a sincere attempt to interest more qualified men in a genuine experience to communicate something Thorp believed important. Certainly, if his experiences were accurately described it is difficult to conceive them arising from the eye itself.

- D. Elwell, Worchester 26 September, 1957

Additional Notes on "Pin Points in the Sky"

∗ About 70% of those participants attending workshops held by the Orgone Biophysical Research Laboratory over the years have been able to make clear visual observations of the atmospheric orgone energy units. Many of the workshop participants had seen these phenomena for years. This compares to our own rough estimate that about 50% of the general population have seen the orgone units, but did not necessarily know what they were.

∗ From one orgone therapist, we learned about a woman who had been on anti-psychotic drugs for many years, as prescribed by a psychiatrist. Her problem? When she was young, she told her parents that she could see "pin points of light in the sky". They promptly sent her to a doctor, who diagnosed her as having "hallucinations", and prescribed the drugs.

∗ Astronauts have seen the orgone energy units while in space, where the partial vacuum and rocket electronics apparently excite them into a more strongly glowing incandescence. We recall John Glenn's orbital trip, in which he saw "swarms" of luminous dots that danced and played around the windows of the space capsule, as it hurtled along at thousands of miles per hour. This event was brilliantly portrayed in the movie "The Right Stuff". Other astronauts have seen bright pin-point flashes light while in space. These reports were either ignored, or dismissed as "cosmic rays" affecting the optical nerve.

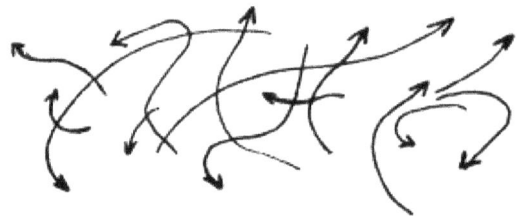

Obesrvable luminating orgone units pulse and randomly move through the sky, with lifetimes of about one second.

Dear Pulse:

I watch in my soul as they destroy the beloved rain forests day by day, just as in Louisiana we are destroying our wetlands. I watch and worry as we destroy more and more of Mother Earth every day.

— Robert D. Tucker, Baton Rouge, LA

Dear Mr. Tucker:

The addresses of a number of environmental groups are given with the articles in the Environmental Notes section. We suggest you join one or more of these, to personally work towards preserving the wetlands and rainforests. A personal and collective, social commitment to sound environmental habits, such as recycling and energy conservation, is needed if the situation is ever going to improve. - J.D.

Dear Pulse:

In the Sedona, AZ area, we have what are called "energy vortexes". They are very real to those of us sensitive enough to feel the energy that exists in them. A place nearby named Boynton Canyon is one of many such Vortex Areas. While hiking inside the canyon 4 years ago, I suddenly felt a "tingling" on the side of my head. Turning, I found the tingling sensation moving behind, front, and then on the opposite side of my head. Following the tingling sensation, I left the trail and headed cross-country. Hours later, I found myself at the location in the enclosed photographs, and definitely felt intense energy in this area, like nowhere else in the Sedona region. I later took a camera with me, and took the enclosed pictures. All were taken on a perfectly clear, cloudless day, yet there are what appears to be cirrus-type clouds above the red rocks. All the pictures were taken within a few minutes of each other, standing in the same spot; if taken from a distance, the "clouds" are in the pictures, *yet when taken close-up, they are not visible.* I have been wondering for a long time, after reading the orgone material, that maybe the orgone energy has places where it has a higher concentration than others, if what I have sensed, felt, and photographed is indeed orgone energy.

— Richard Skoglund, Sedona, AZ

Dear Mr. Skoglund:

The classical theory of orographic convection holds that mountains "force air upward" where it cools and condenses water into cloud droplets. However, speeded-up time lapse images of clouds surrounding high mountains sometimes give the impression that the mountains are literally "spraying" the clouds into the atmosphere, like

a water mist from a fountain, or that they are "flaming" clouds away from their tops, like a torch. Clouds also often appear to get "stuck" at certain mountain locations, irrespective of local air currents. These phenomena do not occur all of the time at all mountain locations, but when they do, the impression is unmistakable, and may indeed have an orgone-energetic basis. Reich observed that clouds are regions of a higher energy charge than the surrounding cloud-free atmosphere. If mountains were discharging orgone energy into the atmosphere, it is very possible that the liberated energy would directly participate in the immediate formation and maintenance of clouds near to those same mountain tops. For other reasons we know that mountain tops are highly charged, in that they are preferred locations for lightning strikes;

Photos taken of "clouds" by Richard Skoglund in the Sedona, Arizona area. Orgone energy streaming?

again, this is discernable from time lapse photographic exposures of mountain tops during stormy conditions. These factors do not rule out orographic convection as a cloud-forming factor, but they do point to other possible cloud formation mechanisms that might explain the various observations you have made. - J.D.

Dear Pulse:

I am interested to know if there has been any work done to prove Reich's theories using Kirlian photography. I've heard of studies done to prove shamanic healing methods by photographing the energy coming from healers hands. It would be a simple matter to prove that Reich's devices concentrate life-energy by simply photographing the [effects of] the device. I would appreciate any information you can give me.

— Vin DiTizio, Jr., Staten Island, NY

Dear Mr. DiTizio:

In response to your question, we quote from a book by Dr. Thelma Moss, who has done a great deal of work on the question of Kirlian, life energy, photography:

"Eva [Reich] was encouraging about our [Kirlian photography] research, which she thought might validate some of her father's theses. She said '...you must not use electricity to take your pictures. ... Dr. Reich always insisted that electricity confounded the orgone effects.' We had been hearing exactly that argument of 'confounding' from those few colleagues who were willing to consider the Kirlian effect. 'There's a huge problem here.' I spoke aloud my confusion. 'You see, it's the electricity that takes the [Kirlian] pictures. How can we take pictures without it?' ... After so many complex failures, the answer came simply. Take apart an ordinary, empty, triple film box, and in total darkness put one piece of color film on the bottom of the empty box. Then, directly on the film, put several small objects, like pieces of fruit, flower buds, etc., making sure they do not touch each other. Next, cover the objects with an 'orgone blanket' layers of steel wool and cotton. Then close the box tightly with both lids so that the inside is light proof and leave it for several days. Then remove the film from the box and have it developed at your favorite color lab. Results? Startling. There is an undeniable, undefinable, Eureka! with an unexpected discovery. I felt this more richly than at any time in my life when I first saw that particular image with its wierd splatterings of Bosch-like images in colors like stained-glass windows. " (From *The Body Electric: A Personal Journey into the Mysteries of Parapsychological Research, Bioenergy, and Kirlian Photography"*, by Thelma Moss, Ph.D., J.P. Tarcher, Los Angeles, 1979, p.136-137)

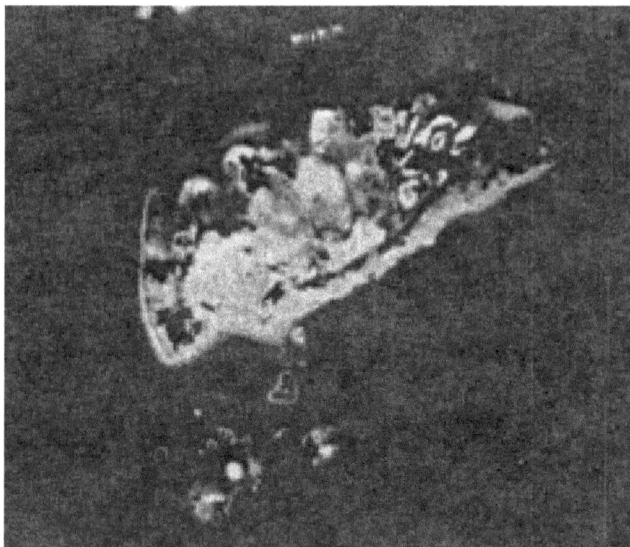

One of Dr. Moss' first non-electrical energy-field photographs (above) of a lemon slice placed on film and covered over with a small orgone blanket.

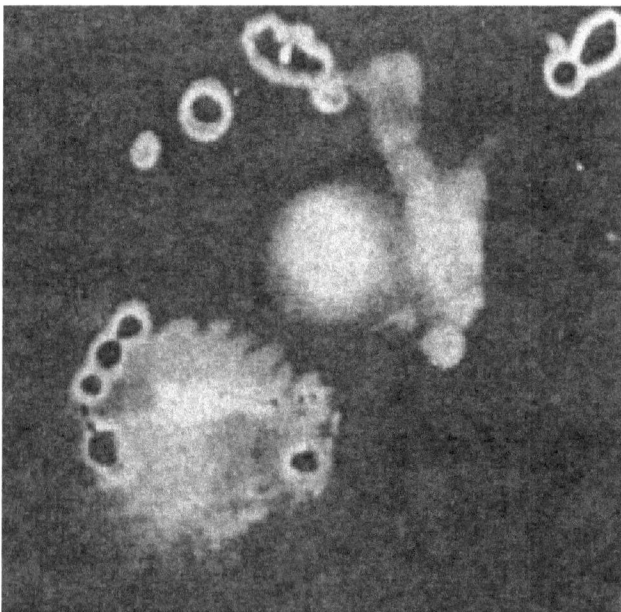

Detail from the "Lemon orgone" picture, showing Bosch-like figures, with nuclei.

Additional details on the non-electrical, orgone-charged energy field photographs of Dr. Moss will appear in a forthcoming issue of the *Pulse.*

A Letter Regarding the Possible Northward Spread of the Sahara Desert into Europe:

Dear Pulse,

For the first time in many years we heard apprehensive comments from Italy that the Sahara will expand northward. Regarding these southwinds coming from the Sahara, we have made very personal experiences: We were at four different times on the Island of Crete in the Mediterranean Sea, and until about 1978 the sky there was a deep blue and crystal clear. In 1981, however, we were there for 3 weeks on our holiday. These 3 weeks were hell. No clouds went by, the light was dull, and the dry, cold, disagreeable wind gave us no chance to find rest anywhere. We felt lost somehow. The cause for this was finally discovered in the southeast of the island. A place named Arvi, much praised in the guidebooks because of its climate, where bananas grow, gave us an impression of a Western ghost town in Arizona. Even dust devils came out to greet us. At first we were too exhausted from the ride to realize all this. We were looking forward to the sea, the sand and sun, and rest. But we did not get any rest. For three days we were puzzled why our heads seemed to be so empty; we had no energy and felt inactive, and we could not think straight. Finally, we felt the inner silence and the void, the slowly creeping death by drying out. At this time there was a strong south wind coming from the Sahara. After we had taken so long to understand this condition, we acted very fast. We took the next bus and rode a few kilometers towards the interior of the country and up a mountain. There it was green and fresh, creeks flowed, and nature (and also we) came to life. This was a very instructive example for us because the transit from the DOR-polluted area to the DOR-free area was so blatant.

What we experience today on the other hand in Middle Europe, is in comparison a gradual transit toward the desert. At the beginning of February of this year we spent a week in the Alps and, although the already mentioned high pressure area predominated, we felt nothing of the DOR at the height of 1800 meters. Only on our way back when we got to the lower regions, it became clear to us in what kind of depressing atmosphere we have to live daily. Even if we get the low pressure areas from the west, our atmosphere is by no way DOR-free. The refreshing affect of the precipitation only holds on shortly; the cold fronts with the typical cumulus formation are mostly missing, and also the typical clear blue of the sky. Even if there are some cumulus clouds, they are heavy loaded with black DOR. Fair weather periods only bring us clear light for one or at best two days, at the latest on the third day the light is dull, the atmosphere sticky. This is the rough outline of our situation since a few years or decades. I personally assume that during the 60s, with all the atom bomb tests,

the atmospheric situation has dramatically worsened. The tendency is, I guess, that the four-season climate will be replaced by two cycles (summer=warm, winter=cold), and both are divided in a dry and moist phase. The only comparison to that is given in the typical monsoon climate. As far as I remember the facts of climatology, the monsoon climates are immediately adjacent to the desert regions. Perhaps they represent a sort of unstable balance, that is to say, a tug-of-war between moist regions and desert.

— Dr. Med. Manfred Fuckert
Centrum fur Orgonomie
~~Memelstrasse 3, Eberbach, 6930~~
West Germany.

Illustration by Beth Cook

Mt. Cook, New Zealand
by Beth Cook

Orgone Biophysical Research Laboratory:

List of Items That Require Additional Donor Support and Funding:

Since the development of the Laboratory in 1978, many projects have been undertaken, particularly regarding the applications and evaluation of the technique of cloudbusting, and in the emergency response to critical drought situations. In recent years, however, this research has developed to the point where *drought prevention* and *desert greening* are real possibilities. Our recent desert experiments in the arid American Southwest, for example, were followed by rains of up to *500% of normal* (see map, page 86). These results fully corroborate Reich's observations and discussions from the early 1950s, that *entire desert regions can be made green and lush*, much as they were in former times, before the genesis of the great desert regions of Earth several thousand years ago. Given that drought and desert spreading are among the most immense and difficult of all contemporary global environmental problems, this research effort is most important. The Laboratory Desert Greening Project, undertaken in the arid lands of the USA, will lay the foundations for desert greening efforts in other lands, where desert spreading and drought are occurring at a faster pace. The financial support necessary for undertaking this next phase of the research is significant, but realistic and achieveable. Here is a list of projects and specific types of laboratory equipment that are required for continued research and operations. Your helpful contributions will be much appreciated.

James DeMeo,Ph.D.
Director of Research

❖ Desert Greening Project: To establish a research faciltiy in the American Southwest
     desert regions for two years minimum. The facility would be devopted to research
     with the Reich cloudbuster, to bring increased moisture to the desert on a consistent
     basis, and green the desert. (Detailed prospectus available for $15): $ 80,000.
❖ OBRL Basic Operations (Includes continuing operations of the Drought Abatement Outreach
     Program, preparation of the *Pulse of the Planet*, continued research on experimental
     orgone biophysics, and various office expenses for 2 years): $ 24,000.
❖ Satellite Image Receiving System, for atmospheric research: $ 12,000.
❖ Apple/Macintosh Computer/Desktop Publishing System, with laser printer, software, misc.: $ 8,000.
❖ Binocular Microscope, for sand/soil bion work, and Reich Blood Test research: $ 8,000.
❖ Two Strip Chart Recorders (2-pen), paper, etc.: $ 5,000.
❖ Geiger-Muller/Scintillation Ratemeter/Counter/Scaler, for detailed study of radioactivity: $ 4,500.
❖ Video Camera/Recorder, Tripod, Accessories, to better document field research operations: $ 3,500.
❖ Laptop Portable Computer, for field work: $ 3,000.
❖ Laboratory Incubator, Autoclave, misc. supplies for bion research: $ 1,200.
❖ Electronic Differential Thermometer, for precision To-T tests: $ 1,200.
❖ National Weather Service weather data, for ongoing analysis of effects of cloudbusting work. $ 1,000.
❖ Filing Cabinets, Office Furniture: $ 800.
❖ Portable Weather Instruments, for field research: $ 500.
❖ Portable precision pH Meter, for acid rains field survey: $ 500.
❖ Precision Light Meter, for field work: $ 400.
                                                                                              ──────────
                                         TWO YEAR FUNDING TOTAL: $153,600.

Donors of $50 - $1000 will automatically receive the *Pulse of the Planet* for one year.
Donors of $1000 or more will automatically receive the *Pulse* for a lifetime.
Call or write for additional details on any of the above items. Your tax-deductible donations may be sent to:
The Orgone Biophysical Research Lab, www.orgonelab.org/donations.htm

Issue Number 1, 1989:

Contains geophysical and climatic data maps for December 1988 through February 1989, plus various environmental notes and research progress reports, and the following major articles:

- Cloudbusting, A New Approach to Drought, by James DeMeo, Ph.D.
- Recent Abnormal Phenomena on Earth and Atomic Power Tests, by Yoshio Kato, Ph.D.
- Response to Martin Gardner's Attack on Reich and Orgone Research in the *Skeptical Inquirer*, by James DeMeo, Ph.D.
- Postscript on the Food and Drug Administration's Evidence Against Wilhelm Reich, by James DeMeo, Ph.D.

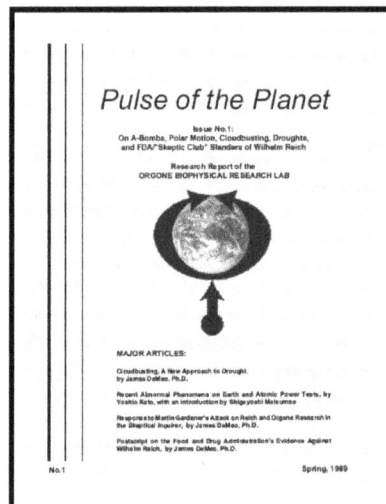

Pulse of the Planet

Issue No.1:
On A-Bombs, Polar Motion, Cloudbusting, Droughts, and FDA/"Skeptic Club" Slanders of Wilhelm Reich

Research Report of the
ORGONE BIOPHYSICAL RESEARCH LAB

MAJOR ARTICLES:

Cloudbusting, A New Approach to Drought, by James DeMeo, Ph.D.

Recent Abnormal Phenomena on Earth and Atomic Power Tests, by Yoshio Kato, with an introduction by Shigeyoshi Matsumae

Response to Martin Gardner's Attack on Reich and Orgone Research in the Skeptical Inquirer, by James DeMeo, Ph.D.

Postscript on the Food and Drug Administration's Evidence Against Wilhelm Reich, by James DeMeo, Ph.D.

No.1   Spring, 1989

Pulse of the Planet, No.1

Scheduled for Forthcoming Issues of the Pulse:

- "The Orgone Accumulator Handbook", exerpts from a new book on the subject by James DeMeo.
- "Dangers of Infant Circumcision", by Marilyn Milos, Director of NOCIRC, the National Organization of Circumcision Information Resource Centers.
- "Body Pleasure and the Origins of Violence", by James Prescott, Editor of the *Truth Seeker*.
- "Female Genital Mutliations" by Fran Hosken, Editor of *Women's International Network News.*
- "The Origins of Armoring in Saharasia", by James DeMeo.
- "Cloudbusting in Israel's Negev Desert", by Rafi Rosen.
- "The Biometer: A New Method for Measurement of Biological Radiations", by Buryl Payne.
- "The Ether-Drift Experiment and the Determination of the Absolute Motion of the Earth", by Dayton Miller, discoverer of the dynamic ether drift.
- More on the unpredicted consequences of underground nuclear bomb testing.
- Book Reviews: the works of Giorgio Piccardi, Louis Kervran, Michel Gauquelin, and more...
- Continuing updates on global environmental, geophysical and climatic factors.
- Continuing updates on new research evaluating and verifying the sex-economic and orgone biophysical discoveries of Wilhelm Reich.

Update: All back issues of Pulse of the Planet are today available from on-line booksellers, as well as from Natural Energy Works (www.naturalenergyworks.net).

www.ingramcontent.com/pod-product-compliance
Lightning Source LLC
Chambersburg PA
CBHW081420270326
41931CB00015B/3346